Mathieu Blanchot

Regio- and Stereoselective Catalytic Addition of Amides to Alkynes

Mathieu Blanchot

Regio- and Stereoselective Catalytic Addition of Amides to Alkynes

Synthesis of the enamide substructure in the most atom-economic and environmentally friendly way

Südwestdeutscher Verlag für Hochschulschriften

Impressum/Imprint (nur für Deutschland/ only for Germany)
Bibliografische Information der Deutschen Nationalbibliothek: Die Deutsche Nationalbibliothek verzeichnet diese Publikation in der Deutschen Nationalbibliografie; detaillierte bibliografische Daten sind im Internet über http://dnb.d-nb.de abrufbar.
Alle in diesem Buch genannten Marken und Produktnamen unterliegen warenzeichen-, marken- oder patentrechtlichem Schutz bzw. sind Warenzeichen oder eingetragene Warenzeichen der jeweiligen Inhaber. Die Wiedergabe von Marken, Produktnamen, Gebrauchsnamen, Handelsnamen, Warenbezeichnungen u.s.w. in diesem Werk berechtigt auch ohne besondere Kennzeichnung nicht zu der Annahme, dass solche Namen im Sinne der Warenzeichen- und Markenschutzgesetzgebung als frei zu betrachten wären und daher von jedermann benutzt werden dürften.

Verlag: Südwestdeutscher Verlag für Hochschulschriften Aktiengesellschaft & Co. KG
Dudweiler Landstr. 99, 66123 Saarbrücken, Deutschland
Telefon +49 681 37 20 271-1, Telefax +49 681 37 20 271-0, Email: info@svh-verlag.de
Zugl.: Kaiserslautern, TU, Diss., 2009

Herstellung in Deutschland:
Schaltungsdienst Lange o.H.G., Berlin
Books on Demand GmbH, Norderstedt
Reha GmbH, Saarbrücken
Amazon Distribution GmbH, Leipzig
ISBN: 978-3-8381-1112-4

Imprint (only for USA, GB)
Bibliographic information published by the Deutsche Nationalbibliothek: The Deutsche Nationalbibliothek lists this publication in the Deutsche Nationalbibliografie; detailed bibliographic data are available in the Internet at http://dnb.d-nb.de.
Any brand names and product names mentioned in this book are subject to trademark, brand or patent protection and are trademarks or registered trademarks of their respective holders. The use of brand names, product names, common names, trade names, product descriptions etc. even without a particular marking in this works is in no way to be construed to mean that such names may be regarded as unrestricted in respect of trademark and brand protection legislation and could thus be used by anyone.

Publisher:
Südwestdeutscher Verlag für Hochschulschriften Aktiengesellschaft & Co. KG
Dudweiler Landstr. 99, 66123 Saarbrücken, Germany
Phone +49 681 37 20 271-1, Fax +49 681 37 20 271-0, Email: info@svh-verlag.de

Copyright © 2009 by the author and Südwestdeutscher Verlag für Hochschulschriften Aktiengesellschaft & Co. KG and licensors
All rights reserved. Saarbrücken 2009

Printed in the U.S.A.
Printed in the U.K. by (see last page)
ISBN: 978-3-8381-1112-4

Abbreviations

2-Fur	*ortho*-furyl
Ac	acetyl
acac	acetylacetonate
Ar	aryl
binol	1,1'-bi-2-naphthol
Bn	benzyl
Cat.	catalyst
cod/COD	1,5-cyclooctadiene
cot/COT	1,3,5,7-cyclooctatetraene
Cp	cyclopentadienyl
Cy	cyclohexyl
dcpe	bis(diphenylphosphino)ethane
dcypb	bis(dicyclohexylphosphino)butane
dcypm	bis(dicyclohexylphosphino)methane
DMA	*N,N*-dimethylacetamide
DMAP	dimethylaminopyridine
DMF	dimethylformamide
dmfm	dimethyl fumarate
DMSO	dimethylsulfoxide
dppb	bis(diphenylphosphino)butane
Eq.	equivalent
Et	ethyl
GC	gas chromatography
HPLC	high pressure liquid chromatography
^{i}Pr/*i*-Pr	*iso*-propyl
L	ligand
M	metal
Me	methyl
met	methallyl
MS	mass spectroscopy
MW	microwave heating
^{n}Bu/*n*-Bu	*normal*-butyl

Abbreviations

^{n}Hex/n-Hex	*normal*-hexyl
NMP	*N*-methylpyrrolidinone
NMR	nuclear magnetique resonance
^{n}Pent/n-Pent	*normal*-pentyl
^{n}Pr/n-Pr	*normal*-propyl
Nu	nucleophile
^{n}Und/n-Und	*normal*-undecane
Ph	phenyl
ppm	parts per million
temp.	temperature
Tf	trifluoromethane sulfonyl
THF	tetrahydrofurane
Tp	tris(pyrazolyl)borate
Ts	4-toluenesulfonyl
X	halide or pseudohalide

Table of contents

Abbreviations i
Table of contents iii
Product numbering iv
Publications iv
1   General part 1
   1.1   Introduction 1
   1.2   Enamides 5
   1.3   Addition of hydrogen bonded nucleophiles to alkynes 14
   1.4   Project aim 46
2   Results and discussion 48
   2.1   Addition of imides to alkynes 48
   2.2   Addition of primary amides to alkynes 64
   2.3   Second generation catalyst for the addition of secondary amides to alkynes 82
   2.4   Mechanistic studies of the hydroamidation of alkynes 99
   2.5   Addition of thioamides to alkynes 104
   2.6   Markovnikov addition of secondary amides to alkynes 112
3   Summary and future aspects 121
4   Acknowledgements 125
5   Experimental part 127
   5.1   General information 127
   5.2   Addition of imides to alkynes 132
   5.3   Addition of primary amides to alkynes 147
   5.4   Second generation catalyst for the addition of secondary amides to alkynes. 167
   5.5   Addition of thioamides to alkynes 181
   5.6   Markovnikov addition of secondary amides to alkynes 194
   5.7   Procedure for the mechanistic studies 195
6   Literature 209

Product numbering

For practicality reasons, the numbering of some compounds is different throughout the dissertation. For example, 1-hexyne appears with the number **2a** in chapter **2.1**, but also with the numbers **8a**, **13a** and **34a** in chapters **2.2**, **2.3** and **2.5** respectively.

Publications

The results of this work were published in the following publications:

L. J. Gooßen, M. Blanchot, K. S. M. Salih, K. Gooßen, *Synthesis,* **2009**, *13*, 2283-2288: *Ruthenium-Catalyzed Addition of Primary Amides to Alkynes: A Stereoselective Synthesis of Secondary Enamides.*

L. J. Gooßen, M. Blanchot, K. S. M. Salih, *Synfacts,* **2009**, *1*, 0085-0085, *12*: *Diastereoselective Synthesis of Secondary Enamides.*

L. J. Gooßen, K. S. M. Salih, M. Blanchot, *Angew. Chem., Int. Ed.* **2008**, *47*, 8492-8495: *Synthesis of Secondary Enamides by Ruthenium-Catalyzed Selective Addition of Amides to Terminal Alkynes.*

L. J. Gooßen, M. Blanchot, K. S. M. Salih, *Org. Lett.* **2008**, *10*, 4497-4499: *Ruthenium-Catalyzed Stereoselective anti-Markovnikov-Addition of Thioamides to Alkynes.*

L. J. Gooßen, M. Arndt, M. Blanchot, F. Rudolphi, *Adv. Synth. Catal.* **2008**, *17*, 2701-2707: *A Practical and Effective Ruthenium Trichloride-Based Protocol for the Regio- and Stereoselective Catalytic Hydroamidation of Terminal Alkynes.*

L. J. Gooßen, K. Gooßen, N. Rodríguez, M. Blanchot, C. Linder, B. Zimmermann, *Pure Appl. Chem.* **2008**, *80*, 1725-1731: *New Catalytic Transformations of Carboxylic Acids: New Catalytic Transformations of Carboxylic Acids.*

L. J. Gooßen, M. Blanchot, C. Brinkmann, K. Gooßen, R. Karch, A. Rivas-Nass, *J. Org. Chem.* **2006**, *71*, 9506-9509: *Ru-Catalyzed Stereoselective Addition of Imides to Alkynes.*

1 General part

1.1 Introduction

"Development that meets the needs of the present without compromising the ability of future generations to meet their own needs".[1] This is the definition for "Sustainable Development", a science in which "Green Chemistry" represents a major component. The term "Green Chemistry" was coined by Anastas[2] of the US Environmental Protection Agency (EPA) in 1993. This does not mean that research on green chemistry did not exist before the early 1990s, merely that it did not have the name. But what does "Green Chemistry" mean?

Nowadays, chemistry is a science so far developed that it is possible to prepare any desired organic compound as long as it is sufficiently stable and its structure can be drawn. Utilization of this capability is most often found in agrochemical, fine chemical or pharmaceutical industries. For instance, certain pharmaceuticals have annual sales exceeding one billion US dollars per years (blockbusters). It is therefore essential to strive for the ideal reaction which produces only the desired compound and minimizes the formation of waste in a simple, safe and environmentally acceptable process. Based on this fact a new chemical philosophy called "Green Chemistry" has emerged almost twenty years ago.

In order to explain what this new philosophy means, Paul Anastas came out with "the twelve principles of *Green Chemistry*".[2]

1) It is better to prevent waste than to treat or clean up waste after it is formed.

2) Synthetic methods should be designed to maximize the incorporation of all materials used in the process into the final product.

3) Wherever practicable, synthetic methodologies should be designed to use and generate substances that possess little or no toxicity to human health and the environment.

4) Chemical products should be designed to preserve efficacy of function while reducing toxicity.

5) The use of auxiliary substances (e.g. solvents, separation agents, and so forth) should be made unnecessary wherever possible and innocuous when used.

6) Energy requirements should be recognized for their environmental and economic impacts and should be minimized. Synthetic methods should be conducted at ambient temperature and pressure.

7) A raw material or feedstock should be renewable rather than depleting wherever technically and economically practicable.

8) Unnecessary derivatization (blocking group, protection/deprotection, temporary modification of physical/chemical processes) should be avoided whenever possible.

9) Catalytic reagents (as selective as possible) are superior to stoichiometric reagents.

10) Chemical products should be designed so that at the end of their function they do not persist in the environment and break down into innocuous degradation products.

11) Analytical methodologies need to be developed further to allow for real-time in-process monitoring and control before the formation of hazardous substances.

12) Substances and the form of a substance used in a chemical process should be chosen so as to minimize the potential for chemical accidents, including releases, explosions, and fires.

Under ideal reaction conditions, substance **A** reacts with substance **B** to produce the desired product (**C**) (equation 1).

1) $$A + B \longrightarrow C$$

But in reality, chemical reactions produce waste (**W**), need temperature (Δ) or pressure (**P**) and use organic solvents (**S**), which could be flammable, volatile, and toxic, with resultant potential threat to both the environment and human health (equation 2).

2) $$A + B \xrightarrow{\Delta, S, P} C + W$$

Nowadays, the use of catalysts represents probably the best way to get close to this utopian ideal reaction. It results in remarkable acceleration of some reactions and can also allow selective action. An ideal catalyst alters the speed of a chemical reaction by its presence while not being consumed itself. In modern chemistry, approximately 80% of all chemicals are produced while in contact with a catalyst in one or more process steps.

A good example to illustrate this idea is the synthesis of hydroquinone (Scheme 1). It was traditionally produced by oxidation of aniline with stoichiometric amounts of manganese dioxide to give benzoquinone. Reduction of the latter by iron and hydrochloric acid gave access to the desired hydroquinone (Béchamp reduction). Aniline was previously derived from benzene *via* nitration and Béchamp reduction. The overall process generated more than 10 kg of inorganic salts per kg of hydroquinone. A more modern route that has now replaced this old process involves autoxidation of *p*-di-isopropylbenzene (produced by Friedel-Crafts alkylation of benzene), followed by acid-catalyzed rearrangement of the bis-hydroperoxide, producing less than 1 kg of inorganic salts per kg of hydroquinone.

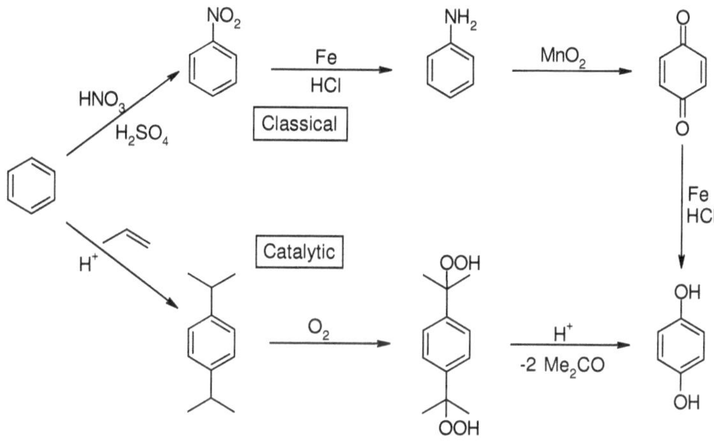

Scheme 1: two routes for the synthesis of hydroquinone.

Another example of an atom-economic reaction is the catalytic addition of hydrogen bonded nucleophiles to alkynes (Scheme 2). Because of the high importance of the products resulting from this kind of transformations, the catalytic additions of Nu-H substrates across C-C multiple bonds have been the focus of many studies both from the academic and industrial world.

However, one particular class of nucleophiles has not been yet studied for this transformation, namely the amides. This remarkable neglect is really very surprising considering the enormous potential of this reaction from a synthetic point of view.

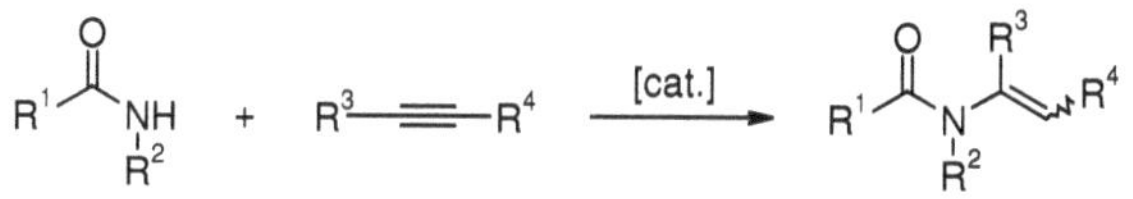

Scheme 2: addition of amides to alkynes.

Introduction

Thus, it appeared essential for us to develop a catalyst system which would allow the direct synthesis of the enamide moiety by addition of an amide to an alkyne.

The aim of this thesis was to achieve the synthesis of the enamide substructure in the most atom-economic and environmentally friendly way. However, as can bee seen in the figure below, many challenges have to be surmounted before a reaction of this type can have broad industrial synthetic applications.

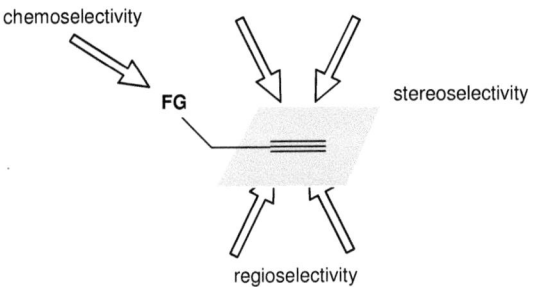

Figure 1: chemo-, stereo- and regiocontrol of the addition of nucleophile across C-C multiple bond.

Since amides are weak nucleophiles, a non-catalytic direct addition of amides to the triple bond of an alkyne is not favored, thus the catalyst system has to overcome first the problem of chemoselectivity so that only attacks on the triple bond will take place and not on the possible functional groups.

Then, the regioselectivity of the reaction has to be controlled in such a way that the product resulting from either a Markovnikov or an *anti*-Markovnikov attack will be formed selectively.

Finally, if the *anti*-Markovnikov product is desired, a catalytic system able to favor only one of the stereoisomers has to be found.

Thus, a perfectly optimized catalytic system must be designed, which should be able to control all these parameters, so that only one product will be formed selectively. This catalytic system should also be highly efficient in terms of turnover number so that the amount of waste products could be reduced to a minimum, according to the atom-economy principles.

1.2 Enamides

1.2.1 Enamides as substructure in natural products

The enamide or *N*-vinylamide functional group is a substructure often encountered in the literature. The enamide linkage represents a fragile structural element that is prominently featured in a wide range of natural products with promising biological activities. Several fungicides,[3] metabolic drugs,[4] and functional materials [5] contain enamides as key structural elements (Scheme 3).

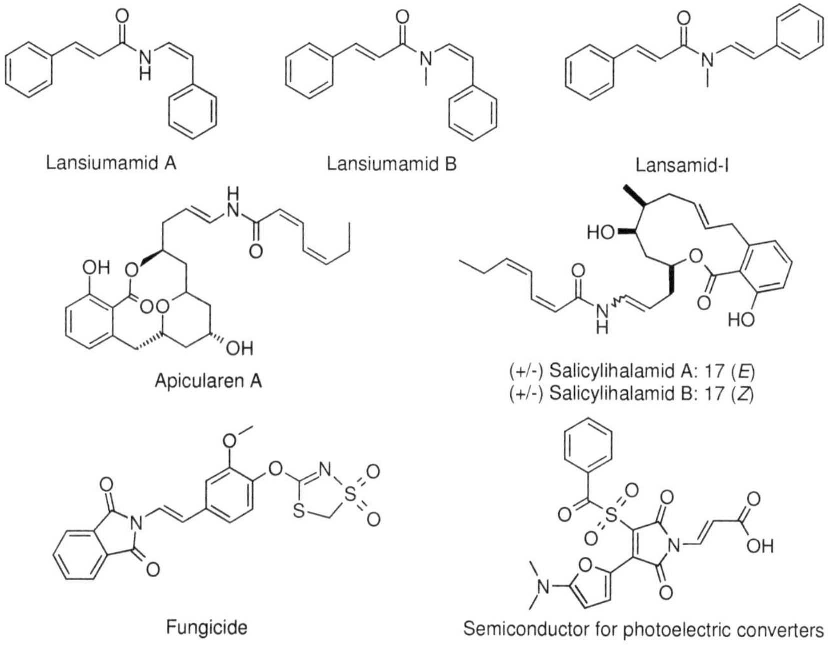

Scheme 3: enamides in bioactive or functional molecule.

For example, the leaves and fruits of *Clausena lansium* have been used as folk medicine in Southern China to treat a range of ailments including asthma and viral hepatitis. Among other cyclic amides, three major compounds have been isolated from the leaves of *Clausena lansium*: Lansiumamid A, B and Lansamid-I.[6] On the other hand, Apicularen A isolated from a myxobacteria and salicylihalamide A and B present in the marine sponge *Halicona* show high activity against human tumor cells.[5]

1.2.2 Enamides as substrates

Enamides are also versatile synthetic intermediates that can serve as substrates for polymerizations, [4+2]-cycloadditions,[7] cross-coupling reactions,[8] Heck olefinations,[9] enantioselective additions [10] or asymmetric hydrogenations.[11] Poly(N-vinyl-2-pyrrolidinone) (PVP) for example is probably the best characterized and most widely studied N-vinyl polymer. Some of the numerous applications of the enamide moiety in organic synthesis are depicted in scheme below.

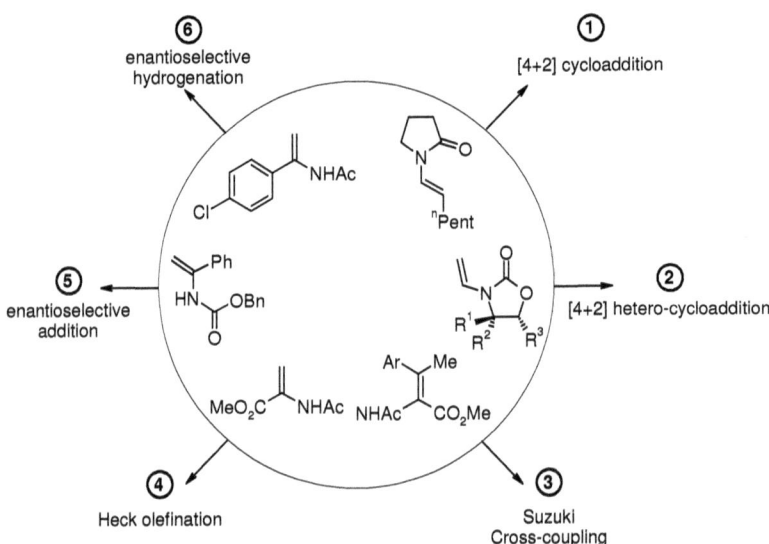

Scheme 4: application of the enamide moiety in organic synthesis.

Numerous heterocycles can be synthesized by [4+2]-cycloadditions using enamides and aromatic imines (Scheme 5)[7a] or Michael-Acceptors (Scheme 6)[7b] as starting materials.

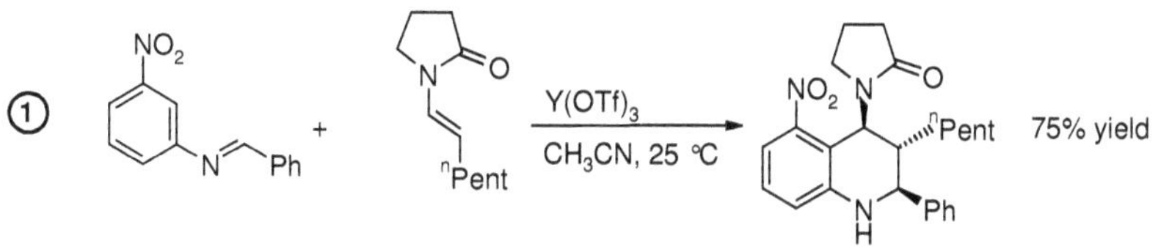

①

Scheme 5: regioselective Lewis acid catalyzed [4+2]-cycloaddition by *Povarov*.

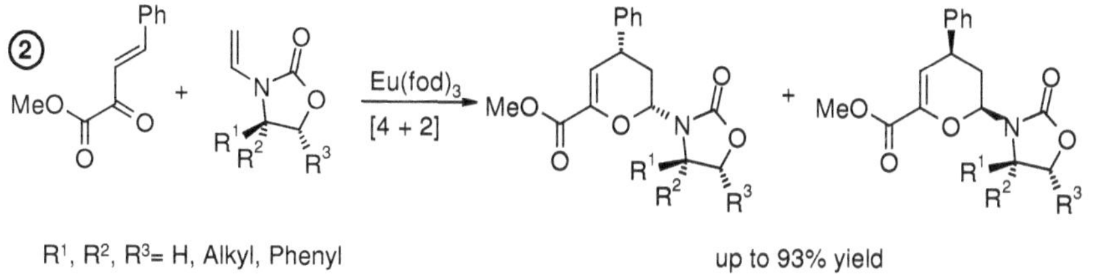

②

R^1, R^2, R^3 = H, Alkyl, Phenyl   up to 93% yield

Scheme 6: synthesis of N-2-deoxyglycosiden by [4 + 2]-heterocycloaddition.

Enamides can also serve as substrates for Suzuki coupling, for example in the synthesis of β-branched α-amino acids (Scheme 7)[8] or for Heck olefination reactions for the synthesis of enantiopure substituted phenylalanines (Scheme 8).[9]

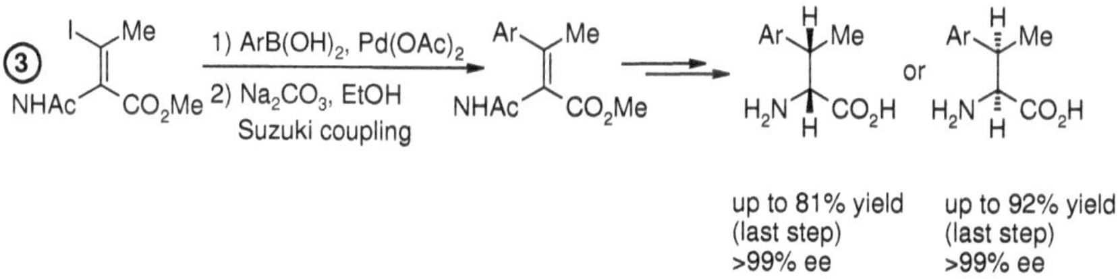

③

up to 81% yield (last step) >99% ee

up to 92% yield (last step) >99% ee

Scheme 7: Suzuki cross-coupling in the synthesis of β-branched α-amino acids.

④

up to 71% yield

Scheme 8: Heck olefination reaction in the synthesis of enantiopure substituted phenylalanines.

Enamides

The copper(II)-catalyzed enantioselective addition of enamides to imines also represents a significant use of the enamide moiety in asymmetric catalysis (Scheme 9).[10]

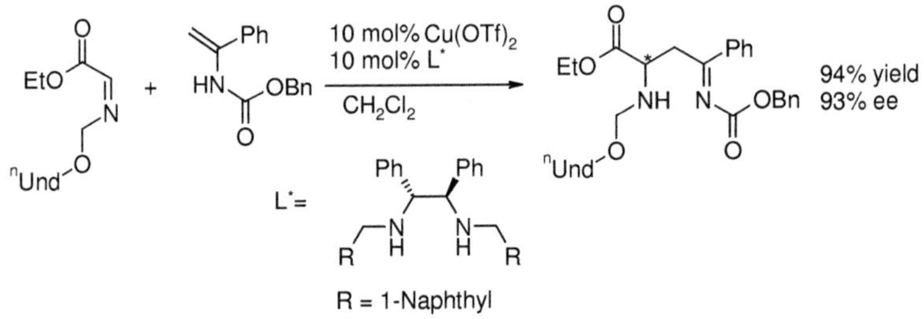

Scheme 9: copper-catalyzed enantioselective addition of enamides to imines.

The enantioselective hydrogenation of enamides catalyzed by chiral rhodium–monodentate phosphite complexes delivers chiral amines in high enantioselectivities (Scheme 10).[11]

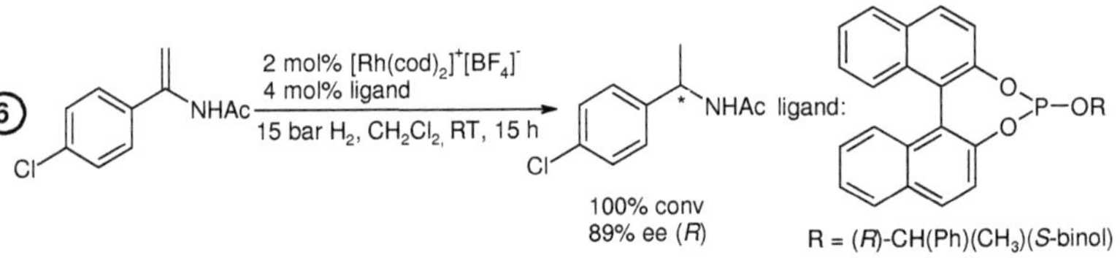

Scheme 10: enantioselective hydrogenation of enamides.

1.2.3 The traditional syntheses of enamides

The regio- and stereoselective construction of enamide substructures represents a major challenge a major challenge for organic chemists and due the high importance of this substructure, there have been numerous attempts to find a way to prepare enamides regio- and stereoselectively. A non-exhaustive list of synthetic ways to prepare enamides is depicted in scheme below.

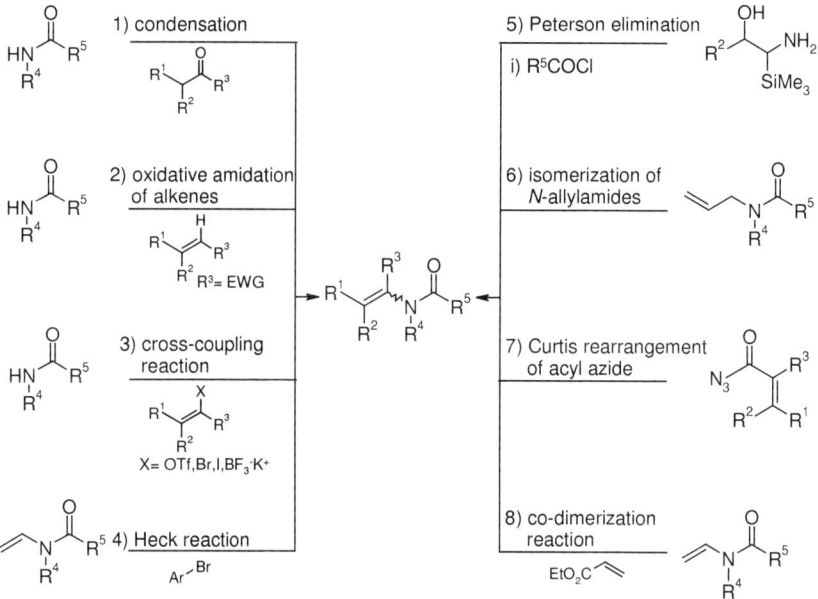

Scheme 11: traditional synthetic pathways for enamides.

Condensation reaction (Scheme 11, equation 1)[12] permits the synthesis of enamides starting from aldehydes and alkynes. This reaction mostly delivers the *E*-isomer or a mixture of the *E/Z*-isomers and often requires the use of strong bases or high temperatures. For example, Zezza and Smith used a condensation reaction for the synthesis of *N*-alkenyl lactams (Scheme 12).[12f]

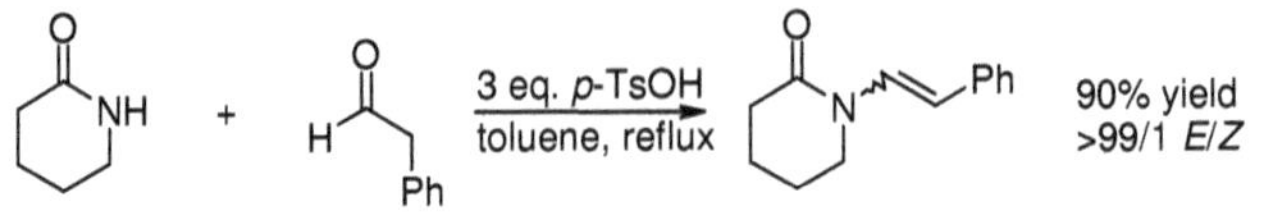

Enamides

Scheme 12: condensation reaction from Zezza and Smith.

The palladium-catalyzed amidation of alkenes (Scheme 11, equation 2)[13] allows the construction of the enamide motif under mild conditions. Starting from alkenes bearing electron withdrawing groups and amides this reaction allows the synthesis of numerous enamides. Nevertheless, the E/Z configuration is highly dependant on the amide employed, and this reaction is strictly limited to a specific class of alkenes. Below, the palladium(II)-catalyzed amidation of alkenes from Murahashi et al. for the synthesis of *(E)-N*-carbamates (Scheme 13).[13a]

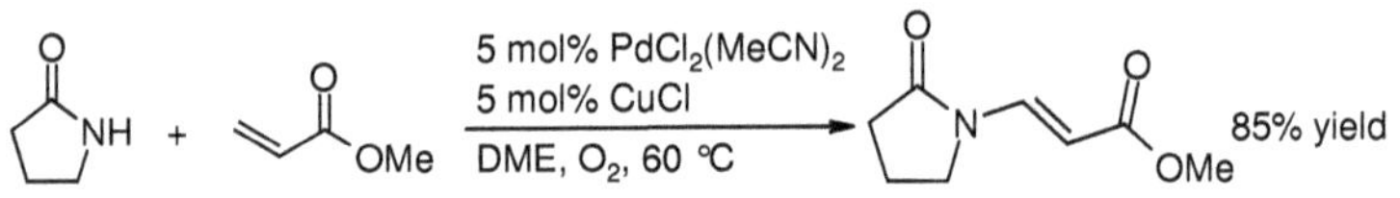

Scheme 13: palladium catalyzed amidation of alkenes from Murahashi.

Palladium or copper catalyzed cross-coupling between amides and vinyl halide, -triflate, -ether or –trifluoroborate salts (Scheme 11, equation 3)[14] allows the stereoselective synthesis of the *E*- or *Z*-isomer in good yields. The geometry of the product is directed by the geometry of the substituted alkene. The major drawback of this reaction is the low availability of the vinyl compounds and/or the ligands employed. A good example for this reaction is the copper catalyzed coupling of vinyl halides with amides promoted by *N,N*-dimethylglycine developed by Ma et al (Scheme 14).[14b]

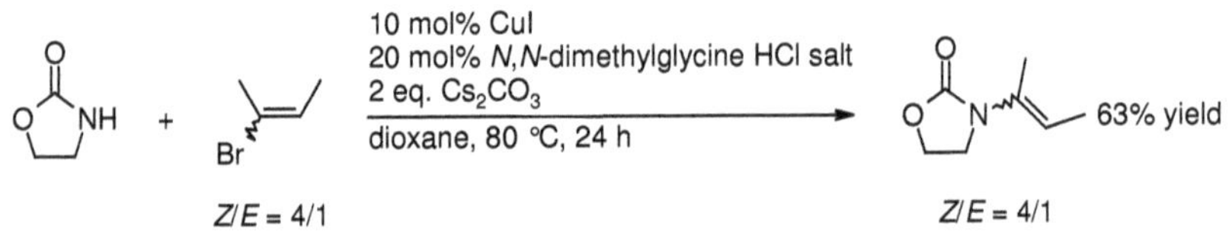

Scheme 14: coupling reaction of vinyl halides with amides under catalysis of CuI and *N,N*-dimethylglycine.

- 10 -

A Heck reaction using vinyl amides and unsaturated halides (Scheme 11, equation 4),[15] such as the reaction developed by Busacca et al. (Scheme 15)[15b] allows the synthesis of enamides. Nevertheless, since one of the starting materials is already an enamide, synthesis of the latter is necessary, moreover, the reaction is limited to unsaturated halides such as aryl halides in most of the cases.

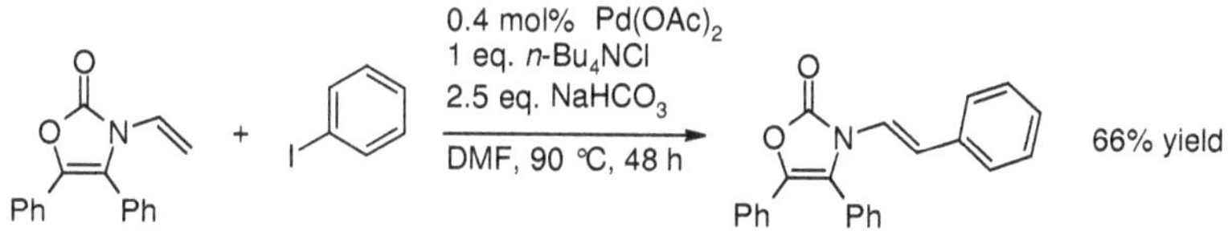

Scheme 15: Heck reaction by Busacca.

After treatment of the α-TMS-β-amino alcohol with the corresponding carboxylic acid chloride, the desired enamide could be obtained by a Peterson elimination in moderate yields. In this case again, the selectivity issue is transferred to the preparation of the α-TMS-β-amino alcohol (Scheme 11, equation 5).[16] Nevertheless, the multistep synthesis of the latter represents the major drawback of this reaction. Fürstner et al. converted vinylsilanes into enamides by a multistep synthesis with a Peterson elimination as a key step reaction (Scheme 16).

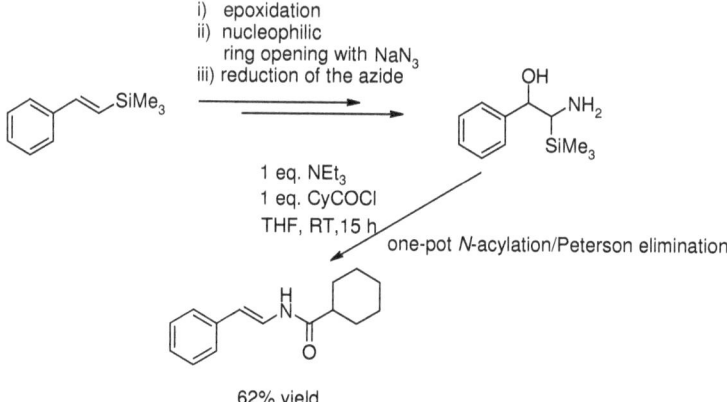

Scheme 16: Peterson elimination as a key step reaction in the synthesis of enamides.

Double bond migration of *N*-allylamides catalyzed by ruthenium or rhodium (Scheme 11, equation 6)[17] is also a convenient method to produce enamides. Drawbacks of the reaction are a poor scope of the reaction limited to *N*-allylamides and no direct availability of the starting materials. An example of this reaction is the ruthenium catalyzed isomerization of *N*-allylamides to corresponding *N*-(1-propenyl)amides from Pigulla et al (Scheme 17).

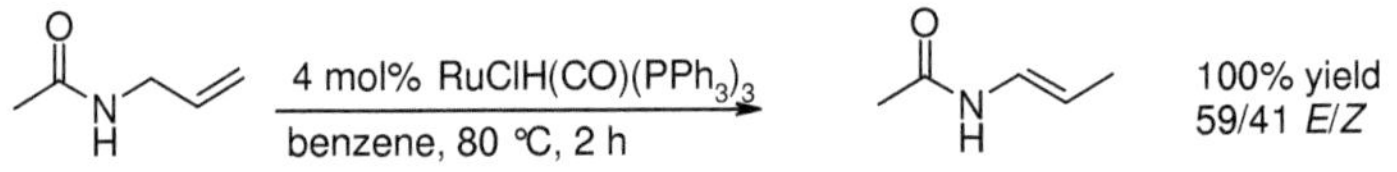

Scheme 17: synthesis of *N*-(1-propenyl)ethanamide by Pigulla et al.

Reaction involving organometallic addition of vinyl Grignards to vinyl isocyanates (Scheme 11, equation 7),[18] which are readily obtained using the Curtius rearrangement of acyl azides, is also a suitable method for preparing enamides. The *E/Z* geometry of the product is directed by the geometry of the vinyl isocyanate, nevertheless, if the *Z*-isomer is the desired product only the reaction carried out at very low temperatures will preserve the *Z*-geometry, otherwise, the *E*-isomer predominate. An interesting application of this method is the synthesis of the natural product Lansamide I developed by Taylor et al. (Scheme 18).[6]

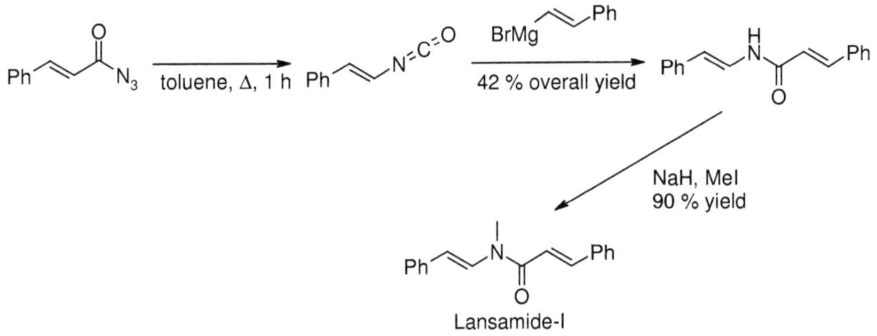

Scheme 18: synthesis of Lansamide-I by Tailor et al.

Recently, Kondo et al. have developed the ruthenium-catalyzed co-oligomerization of *N*-vinylamides with alkenes (Scheme 11, equation 8)[19] which provides *E*-enamides in good yields (Scheme 19). However, the preparation of the *N*-vinylamides is not trivial at all, and represents the same major disadvantage of the reaction.

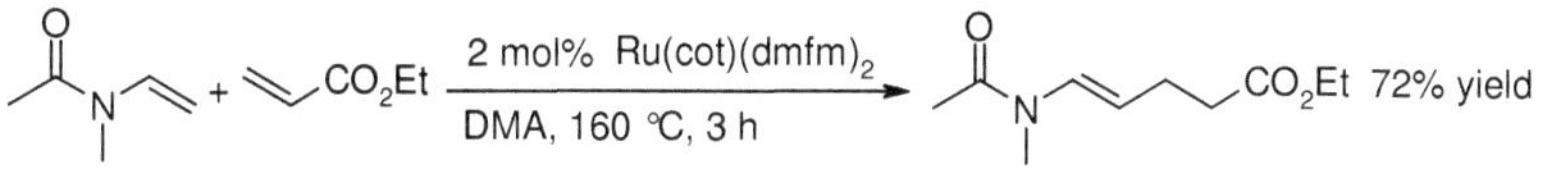

Scheme 19: co-oligomerization of *N*-vinylamides with alkenes.

Nevertheless, none of these methods allows a concise and environmentally benign approach for the synthesis of enamides. Insufficient availability of the starting materials, poor selectivity of the reaction or harsh reaction conditions are the main drawbacks.

A catalytic addition of amides to alkynes would be an ideal synthetic entry to enamides, since it would use readily available starting materials and be inherently atom-economic. Moreover, similar reactions, e.g. addition of hydrogen-bonded nucleophiles such as carboxylates, water, and amines are well established and represented a great source of inspiration for our work.

1.3 Addition of hydrogen bonded nucleophiles to alkynes

The addition of hydrogen bonded nucleophiles (Nu-H) across the carbon-carbon triple bond, catalyzed by transition-metal complexes, is one of the most interesting and intriguing subjects in organic chemistry. Only four types of nucleophiles (carboxylates, water, amines and amides) will be covered in the following chapter; but this kind of chemistry also includes other different types of bonds, such as S-H, Se-H, and P-H.

The atom economy of this process, coupled with minimized amounts of produced waste, make this chemistry far more environmentally friendly than substitution reactions leading to the same products, and explain the keen interest in and the importance of these transformations. Resulting from this reaction, very useful alkenyl organic compounds, such as nitrogen- and oxygen- containing products, can be obtained. Moreover, the intramolecular version of this reaction represents a straightforward way to obtain heterocycles. These transformations allow very mild reaction conditions when catalyzed by transition metals, and proceed with good yields and selectivities if the appropriate catalyst system is employed.

It is likely that the major challenge encountered while performing this kind of transformation is the control of the regio- and stereoselectivity. Indeed, if terminal alkynes are employed, three different products could be obtained. One product as a result of a Markovnikov addition, often described as "Markovnikov product" (Scheme 20, equation 1), and two other products resulting from an *anti*-Markovnikov addition, which differ in their stereochemistry Z or E (Scheme 20, equation 2).

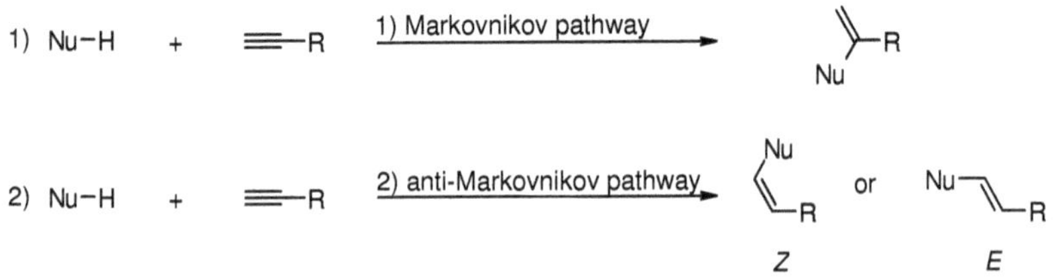

1) Nu−H + ≡−R →(1) Markovnikov pathway→ $\overset{\text{Nu}}{\underset{}{\diagup\!\!\!\diagdown}}$−R

2) Nu−H + ≡−R →(2) anti-Markovnikov pathway→ $\overset{\text{Nu}}{\diagdown}$−R or Nu−$\diagdown$−R

                                                                                                   Z   E

Scheme 20: regio- and stereoselectivity of the addition of Nu-H to terminal alkynes.

The type of transition-metal complex used, the nature of the hydrogen bonded to the nucleophile which can be totally different depending on the electronegativity of the heteroatom, and the electronic effect of the alkynes are parameters that strongly affect the selectivity of those reactions.

Attempts were made to understand the reasons for a reaction to proceed either *via* a Markovnikov or an *anti*-Markovnikov pathway. As depicted below, the types of activation can be divided into 3 categories.

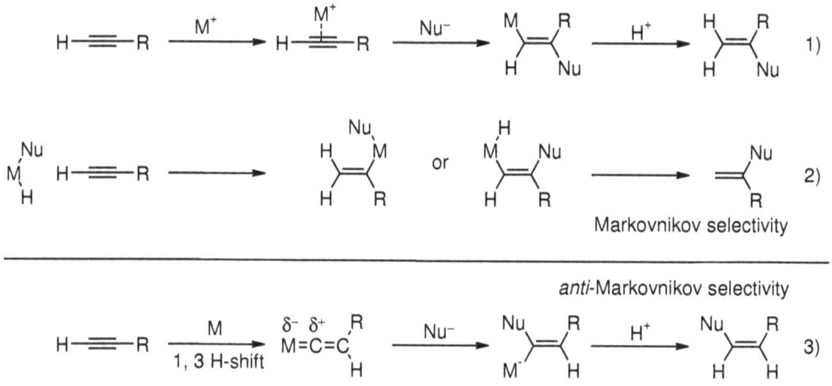

Scheme 21: different modes of activation for the addition of nucleophile to alkynes.

Activation of the carbon-carbon triple bond toward nucleophilic attack can be performed by metals with high Lewis acidity; this alkyne activation type should lead, after addition of the nucleophile, to a product with Markovnikov selectivity (Scheme 21, equation 1). On the other hand, the same selectivity is described if the reaction proceeds *via* insertion of the alkyne moiety into the M-H or M-Nu bond of an activated nucleophile (typically, complex of the type H-M-Nu, Scheme 21, equation 2). To explain the *anti*-Markovnikov selectivity, Dixneuf et al. mentioned the possibility of formation of a ruthenium vinylidene species with an electron-deficient Ru=C carbon site (Scheme 21, equation 3).

There is no doubt that such an approach to the synthesis of many natural products or building blocks represents great advantages, particularly because of the relative readily availability and low costs of the starting materials. The main problem that chemists have to face in these reactions is to find the particular catalytic system which would allow the reaction to proceed regio- and stereoselectively, and also be relatively cheap and stable at the same time. The aim of the following chapter is to summarize some of the most important results in this area.

1.3.1   Addition of water to alkynes

A painting by M. Kucherov of the Chemical Laboratory at the Imperial Forestry Institute in St. Petersburg, Russia at the end of 19th century.[20]

In organic molecules, carbonyl groups are among the most common functionalities. A large number of different product classes have become accessible from this single functionality along with multiple faceted reaction pathways.

A relevant example is the industrial production of acetaldehyde synthesized by acid catalyzed hydration of acetylene using mercury salts.[21] The development of this transformation was a great revelation at that moment but was far from complete. However, this process has been intensively used for the industrial production of acetone for the explosives manufacture during the First World War in Germany. After the conflict and based on the hydration of acetylene, a variety of industrial chemicals have been produced on large scale over many years (Scheme 22).

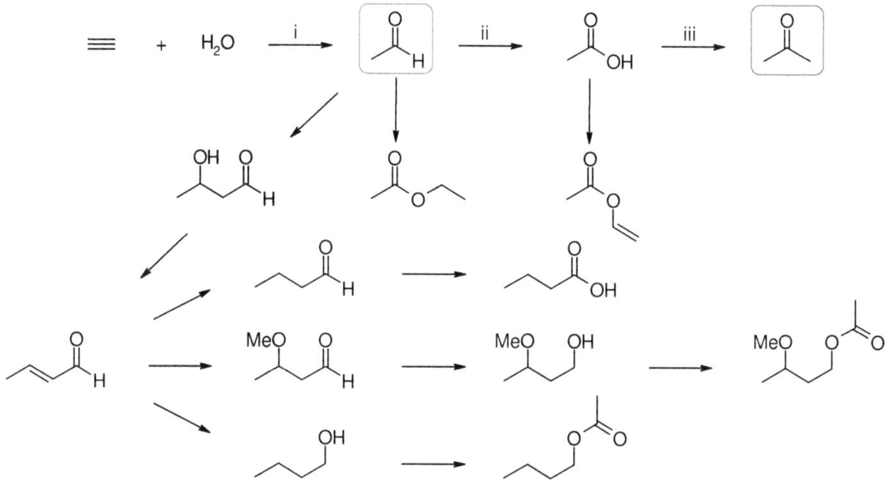

i = hydration of acetylene using mercury (II) salts.
ii = oxydation of acetic acid using oxygen and manganese acetate.
iii = CaCO$_3$ + 2 CH$_3$COOH $\longrightarrow$ Ca(CH$_3$COO)$_2$ + CO$_2$ + H$_2$O
Ca(CH$_3$COO)$_2$ $\xrightarrow{\Delta}$ CaO + CH$_3$(CO)CH$_3$ + CO$_2$

Scheme 22: chemicals accessible from acetaldehyde.

Acid-catalyzed hydration of alkynes has been known for a long time. Marcellin Berthelot, during his work on acetylene (1860), described its hydration in sulfuric acid and believed he had obtained vinyl alcohol,[22] but it turned out that the product synthesized was a mixture of acetaldehyde and crotonaldehyde.[23] Twenty years later, Kucherov described the mercury salt catalyzed hydration of acetylenes and alkynes and first introduced the concept of "hydration".[24] A process developed by Kucherov and using mercury (II) sulfate is nowadays still in use[25] and was employed all over the years in many applications (Scheme 22). However, pollution by toxic mercurial waste had always been a problem of the process, and research aiming to reduce the toxicity of the reaction by replacing mercury by less toxic metals has continued over the years[26] since the addition to C-C triple bond is desirable in terms of atom-economy in forming C-O double bond compounds.

The acid-catalyzed hydration of alkynes proceeds by protonation of the carbon–carbon triple bond, followed by fast addition of water to generate an enol which leads to the corresponding ketone since the reaction is directed by the Markovnikov rule. However, only electron-rich alkynes such as phenyl acetylene react satisfactorily by this method. The reaction of simple

Addition of hydrogen bonded nucleophiles to alkynes

alkynes is usually sluggish and needs co-catalysts, typically mercury (II) salts such as mercury (II) chloride, mercury (II) bromide, mercury (II) acetate or mercury (II) sulfate to enhance the reactivity. The role of the catalyst is to polarize the alkyne in order to facilitate the attack of water. Kucherov discovered the activity of mercury(II) bromide catalyst in the hydration of alkynes and later, succeeded in hydrating propyne to acetone and acetylene to acetaldehyde under mild conditions by the use of dilute sulfuric acid and mercuric salts as catalysts (Scheme 23, equation 1).[24] Carter and Conaway successfully applied the same method to the synthesis of methyl vinyl ketone from vinylacetylene (Scheme 23, equation 2).[27] Finally, Hennion et al. extended the reaction to higher alkynes (Scheme 23, equation 3).[28] Scheme 23 depicts some examples for this transformation.[29] The regiochemistry is indicative of a mercurinium ion intermediate which is in turn opened by nucleophilic attack at the more positively charged carbon; that is, the addition follows the Markovnikov rule. Terminal and disubstituted alkynes could also be converted into ketones with very good results.

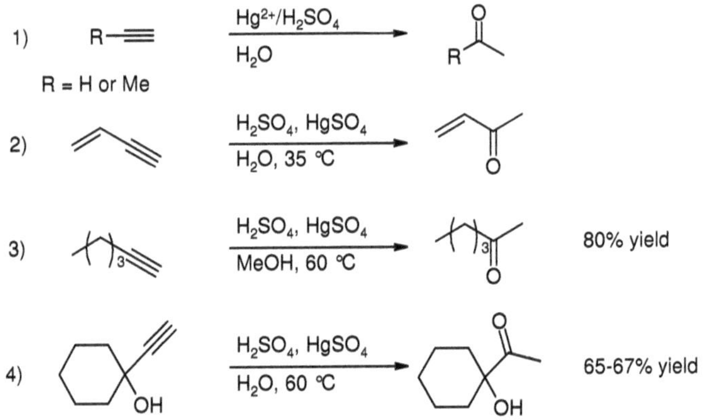

Scheme 23: typical mercury catalyzed hydration of alkynes.

Further studies showed that other metal species such as Ag (I), Pd (II), Rh (III), Pt (II) are also useful for this conversion.[26] Noteworthy are platinum(II)-salts such as Zeise's dimer [{PtCl$_2$(C$_2$H$_4$)}$_2$], that show promising activities for the hydration of non-activated alkynes.[30] The mechanisms of these reactions are identical for all of these metals with subtle differences due to the relative stabilities of the intermediates. Meanwhile several aspects of the reaction are still unclear and only simple mechanisms such as the one depicted in scheme below are described in the literature.

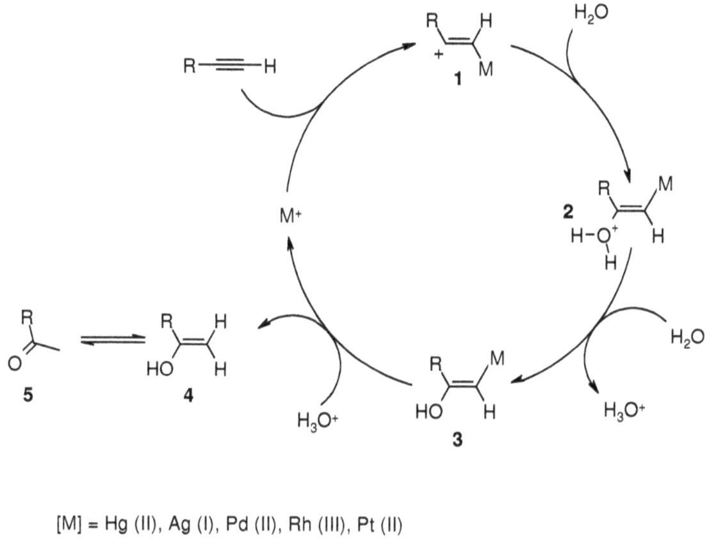

[M] = Hg (II), Ag (I), Pd (II), Rh (III), Pt (II)

Scheme 24: proposed mechanism for the mercury catalyzed hydration reaction.

The alkyne uses a pair of electrons to attack the electrophilic metal ion, yielding a metal-containing vinylic carbocation intermediate (**1**). Nucleophilic attack of water on the carbocation forms a C-O bond and yields a protonated metal-containing enol (**2**). Abstraction of H^+ from the protonated enol by water gives the organometallic compound **3**. Replacement of M^+ by H^+ occurs to give a neutral enol (**4**), which undergoes tautomerization to give the final ketone product (**5**).

Gold has traditionally been regarded as inactive for catalysis, but it has gained interest in the last decades. When applied to the hydration of alkynes, it is able to convert terminal and internal alkynes into the corresponding ketones. The first gold-catalyzed hydration was observed in 1976 using $HAuCl_4$ but it suffered from low yields.[31] A breakthrough in the field was made by using catalysts of the type $[L-Au^+]$ reported by Teles,[32] highly active for the additions of alcohols to alkynes, while they also appear to be very active for the hydration of alkynes simply by using methanol as solvent.[33] Gold ions with a d^{10} electronic configuration, such as $[(PPh)_3AuMe]$, have to be combined with a strong acid such as H_2SO_4 in order to generate the cationic active species that is able to catalyze the hydration of alkynes. Activated and non-activated terminal alkynes as well as internal alkynes are converted to the corresponding ketones in very good yields and selectivities. Although the mechanism of the reaction is not perfectly understood; it could be assimilated to the one described for the gold catalyzed addition of alcohols to alkynes (Scheme 25).[32]

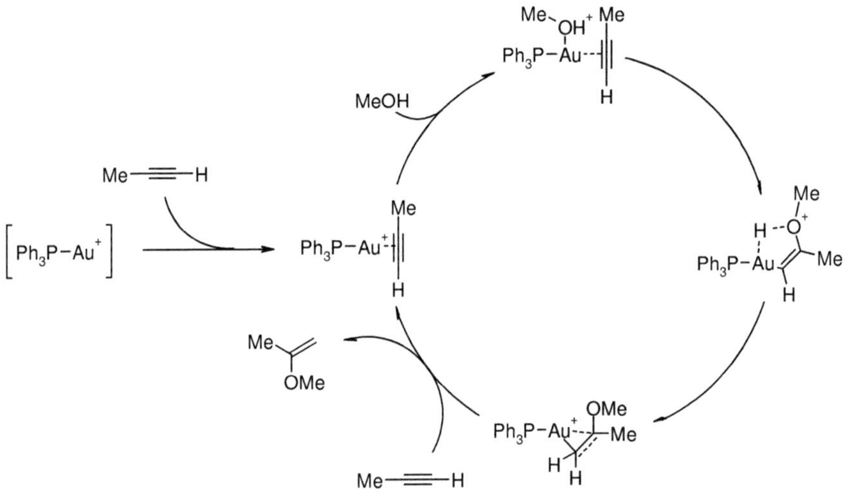

Scheme 25: proposed mechanism for the gold catalyzed addition of alcohols to alkynes.[12]

Unfortunately, all of these reactions reported so far follow the Markovnikov rule and exclusively lead to ketone products in the case of terminal alkynes.

Ruthenium is one of the prominent metals in alkyne chemistry, however when applied to the hydration of alkynes it first appears to only suffer from *in situ* deactivation due to C-C bond cleavage of the terminal alkynes by water assisted by ruthenium [34] with release of alkanes. The mechanism of this reaction described by Bianchini at al. (Scheme 27) highlights the possibility of the formation of a vinylidene complex responsible for this deactivation. However, other catalysts such as [RuCl$_2$(C$_6$H$_6$)(PPh$_3$)] are able to convert aryl or alkyl alkynes to the corresponding ketones as well as aldehydes as side products.[35]

As described above, the ability of ruthenium to form vinylidene intermediates with alkynes leads to a deactivation of the catalyst, but it appears that this peculiarity represented a way to reverse the selectivity of the reaction. In 1998, more than 100 years after the first catalyzed hydration of alkynes, Tokunaga and Wakatsuki developed the first aldehyde synthesis based on catalytic *anti*-Markovnikov hydration of terminal alkynes.[36] This novel hydration of terminal alkynes catalyzed by ruthenium(II) complexes combined with an excess of the appropriate phosphine ligands leads to predominantly *anti*-Markovnikov addition of water and yields aldehydes with only a small amount of ketones (Scheme 26).

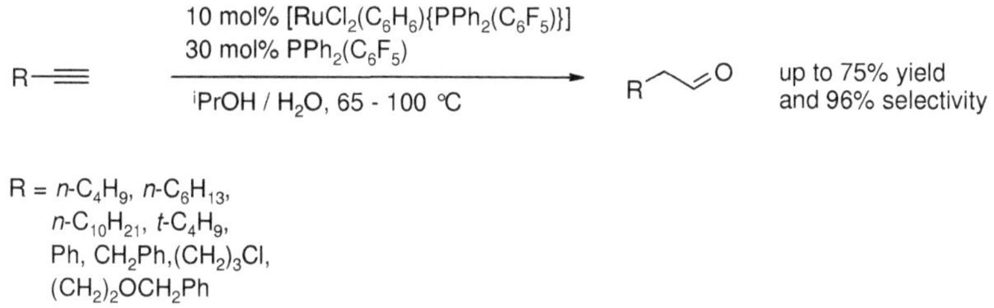

$$R\text{—}\equiv \xrightarrow[\text{iPrOH / H}_2\text{O, 65 - 100 °C}]{\substack{10 \text{ mol}\% \text{ [RuCl}_2(\text{C}_6\text{H}_6)\{\text{PPh}_2(\text{C}_6\text{F}_5)\}] \\ 30 \text{ mol}\% \text{ PPh}_2(\text{C}_6\text{F}_5)}} R\diagdown\!\!\diagup^{\text{O}} \quad \text{up to 75\% yield and 96\% selectivity}$$

R = n-C$_4$H$_9$, n-C$_6$H$_{13}$, n-C$_{10}$H$_{21}$, t-C$_4$H$_9$, Ph, CH$_2$Ph, (CH$_2$)$_3$Cl, (CH$_2$)$_2$OCH$_2$Ph

Scheme 26: first catalytic *anti*-Markovnikov hydration of terminal alkynes.

The mechanism of the reaction is still under investigation but could be summarized as follows (Scheme 27). It is likely that the ruthenium catalyzed *anti*-Markovnikov hydration of alkynes involves a tautomerization of the 1-alkyne complex **7** to the vinylidene complex **8**, generating a metal-acyl species (**10**) after attack of the water oxygen atom. This acyl intermediate would undergo the protonolysis of a carbon-metal bond with release of an aldehyde (**15**).

The C-C bond cleavage or decarbonylation reaction (**13** to **14**) is expected to occur together with the main reaction, formation of aldehyde, since it is a highly favored reaction pathway in the stoichiometric systems reported by Bianchini et al.

On the other hand, the formation of the ketone **9** involves a nucleophilic attack of water on a coordinated alkyne (**7**) with Markovnikov selectivity; this mechanism is similar to the one involving mercury (II).

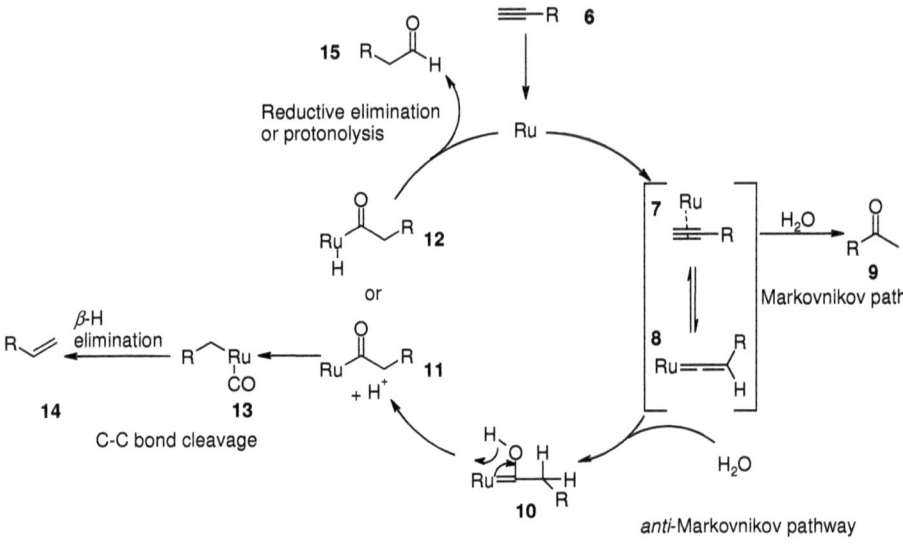

Scheme 27: mechanism of the ruthenium catalyzed hydration of alkynes.

Depending on the type and amount of phosphine used, both *anti*-Markovnikov and Markovnikov products can be accessed regioselectively (Scheme 28). The role of the phosphine is probably to shift the equilibrium between **7** and **8** (Scheme 27).

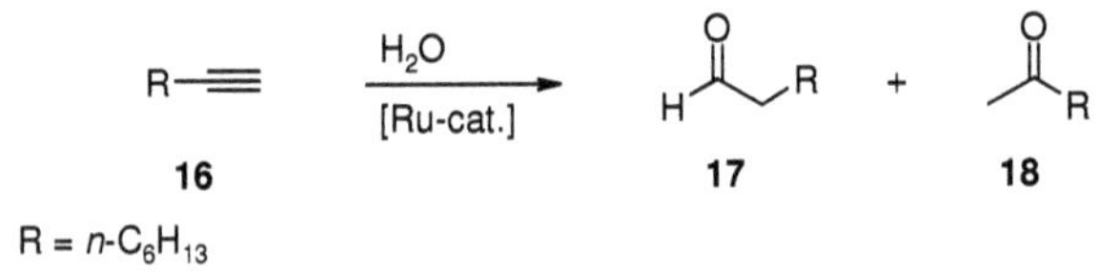

R = *n*-C$_6$H$_{13}$

catalytic system	yield (%)[a] of 17	yield (%)[a] of 18	selectivity 17/18
[RuCl$_2$ (C$_6$H$_6$)(PPh$_3$)][b]	8	66	1/8
[RuCl$_2$ (C$_6$H$_6$)(PPh$_3$)] + 3 PPh$_3$	9	2	5/1
[RuCl$_2$(C$_6$H$_6$){PPh$_2$(C$_6$F$_5$)}] + 3 PPh$_2$(C$_6$F$_5$)[c]	75	4.5	17/1
[RuCl$_2$(C$_6$H$_6$)]$_2$ + 8 P(3-C$_6$H$_4$SO$_3$Na)$_3$	0	6.1	-
[RuCl$_2$(C$_6$H$_6$)]$_2$ + 8 P(3-C$_6$H$_4$SO$_3$Na)$_3$[d]	45	4.5	10/1

Reaction conditions: 10 mol % of the Ru-source and 30 mol % phosphine when needed in 2-propanol at 100 °C for 40 h. a) GLC yields. b) 14 h. c) 12 h and 65 °C. d) in 2-methoxyethanol.

Scheme 28: different phosphine and catalyst combinations for the hydration of 1-octyne.

At present, ruthenium remains the only metal able to catalyze this reaction and the only major improvement was made by Wakatsuki et al. by using a cyclopentadienyl ligand. Complexes of the type [CpRu(PR$_3$) Cl] remarkably increase the rate and the selectivity of the reaction.[36]

Catalytic *anti*-Markovnikov hydration of terminal alkynes is a new reaction with great potential in organic synthesis. Its study has just begun and there is no doubt that fascinating applications will result from this type of chemistry. Moreover, since the riddle of its mechanism is not yet solved, it will be interesting to see whether metals other than ruthenium can catalyze the reaction.

1.3.2 Addition of carboxylic acids to alkynes

Enol esters have been used in two major areas, polymerization and acylation. The simplest enol ester is the key monomer for access to polyvinyl acetate (Scheme 29, equation 1), while enol esters are activated esters and can be used as mild and efficient acylating reagents in organic synthesis: on acylation, they release a neutral and stable ketone (Scheme 29, equation 2).

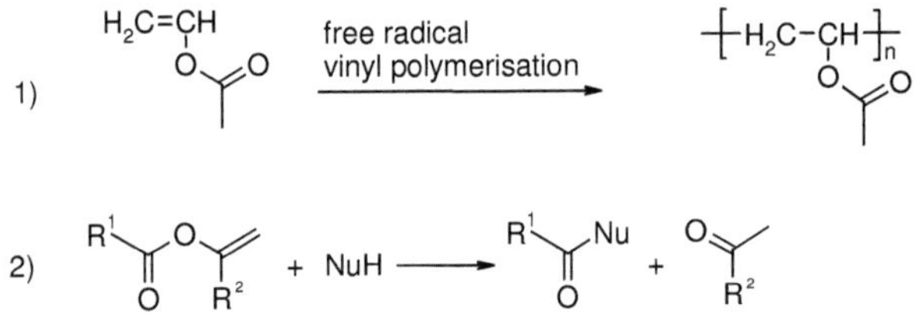

Scheme 29: polymerization and acylation using enol esters.

Treatment of aldehydes or ketones under either acidic or basic conditions with the appropriate acid anhydride or chloride[37] or acetoxylation of olefins promoted by palladium acetate[38] are major methods to obtain enol carboxylates. Among these methods, the addition of carboxylic acids to alkynes, in which mercury catalysts are very useful catalysts, is probably the most elegant way to synthesize enol esters.

In 1892, Béhal and Desgrez reported a reaction between 1-heptyne and acetic acid, when these were heated at 280 °C for twenty-four hours in a sealed tube.[39] The only product identified after completion of the reaction was the 1-methylhexyl acetate **19** which they assumed was formed by the degradation of the α-vinyl acetate **20**, which they did not isolate (Scheme 30).

In 1933, Nieuwland et al. confirmed these observations.[39] Performing the addition of acetic acid to 1-heptyne, they obtained 34% of the corresponding α-vinyl acetate **20** and the methyl ketone **19** as a side product of the reaction (3%). The use of mercury (II) oxide in combination with boron fluoride catalyzes the reaction and permits the use of a lower temperature (40 °C) which probably slows down the degradation of compound **20** to **19**.

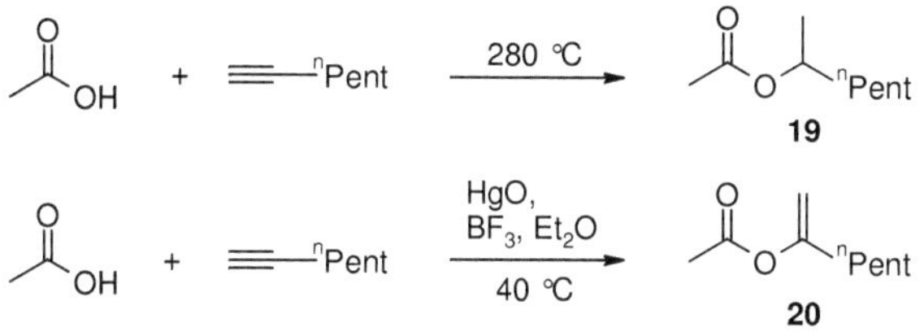

$$\underset{OH}{\overset{O}{\bigwedge}} + \equiv\text{---}^{n}\text{Pent} \xrightarrow{280\ °C} \underset{\substack{\\19}}{\overset{O}{\bigwedge}_{O}\overset{}{\bigwedge}^{n}\text{Pent}}$$

$$\underset{OH}{\overset{O}{\bigwedge}} + \equiv\text{---}^{n}\text{Pent} \xrightarrow[40\ °C]{\text{HgO, BF}_3,\text{ Et}_2\text{O}} \underset{\substack{\\20}}{\overset{O}{\bigwedge}_{O}\overset{}{\bigwedge}^{n}\text{Pent}}$$

Scheme 30: reaction of 1-heptyne with acetic acid.

Since vinyl acetate is an important raw material for synthetic resins, addition of acetic acid to acetylene has received the major share of attention. Until 1970, addition of acetic acid to acetylene in the presence of mercury or zinc acetate on activated charcoal was one of the major industrial processes to produce vinyl acetate. Even if the reaction conditions are harsh, it is possible to prepare vinyl stearate and higher vinyl esters, or hindered esters by this method (Scheme 31).[40]

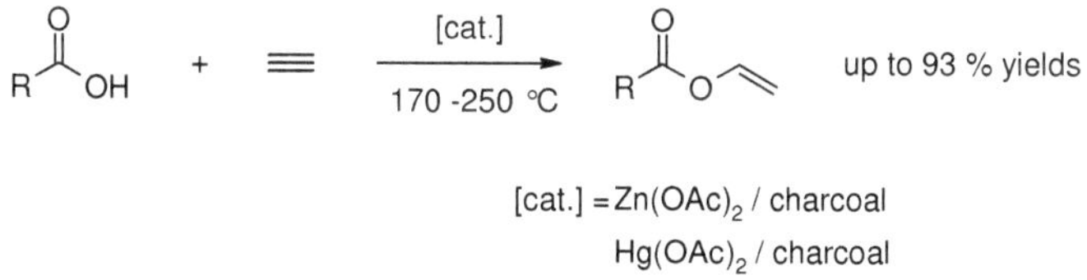

$$R\overset{O}{\underset{OH}{\bigwedge}} + \equiv \xrightarrow[170\text{ -}250\ °C]{[\text{cat.}]} R\overset{O}{\underset{}{\bigwedge}}_{O}\diagdown \quad \text{up to 93 \% yields}$$

$$[\text{cat.}] = \text{Zn(OAc)}_2 / \text{charcoal}$$
$$\text{Hg(OAc)}_2 / \text{charcoal}$$

Scheme 31: industrial preparation of vinyl acetate.

Nevertheless, the scope of the reaction is very limited because of the high temperatures used. Moreover, since the mercury salts are very toxic, development of safer catalysts was required. Besides, similarly to the hydration of alkynes, the mercury catalyzed addition of carboxylic acids to alkynes leads to the Markovnikov product and intensive studies were made to develop suitable methods to obtain the *anti*-Markovnikov product.

No major improvement was made in this area until Rotem and Shvo in 1983 described the first ruthenium catalyzed addition of carboxylic acids to alkynes.[41] They showed that $\text{Ru}_3(\text{CO})_{12}$ is able to promote the addition of carboxylic acids to diphenylacetylene at 145 °C in toluene. Subsequently Watanabe et al. found and reported a catalyst formed *in situ* from bis(cod)ruthenium, a trialkyl phosphine (*n*-Bu$_3$P or PCy$_3$) and maleic anhydride, able to catalyze the Markovnikov addition of carboxylic acids to terminal alkynes in high yields and selectivities (Scheme 32).[42]

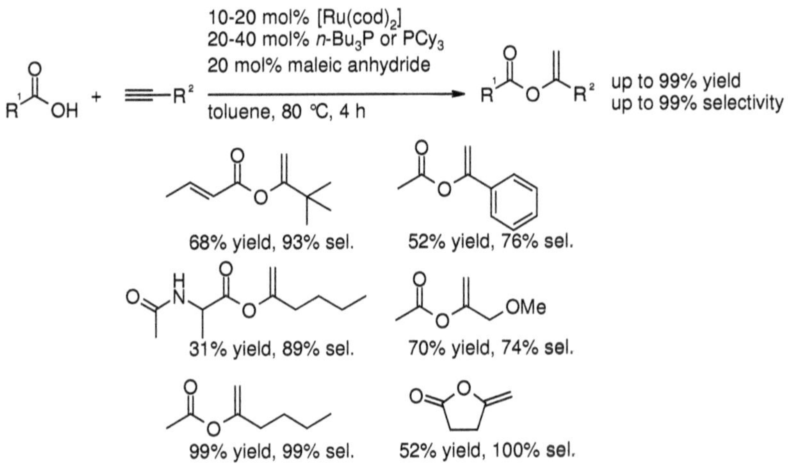

Scheme 32: ruthenium-catalyzed Markovnikov addition of carboxylic acids to alkynes.

Almost at the same time, Dixneuf et al. made an important contribution to this field by describing [RuCl$_2$(p-cymene)(PPh$_3$)] and [{Ru(O$_2$CH)(CO)$_2$(PPh$_3$)}$_2$] as similarly widely applicable and highly Markovnikov-selective catalysts.[43] Even N-protected amino acids and peptides can be converted smoothly into vinyl esters with these catalysts. The authors also developed a complementary system from bis(methallyl)ruthenium and the chelating phosphine 1,4-bis(diphenylphosphino)butane (dppb), with which the (Z)-*anti*-Markovnikov products are selectively accessible for the first time.[43]

Furthermore, based on the previous kinetic investigations made by the group of Watanabe who showed that the addition of carboxylic acid is rate determining in the Markonikov-selective protocol,[42] Gooßen et al. developed a simple catalyst system in which catalytic amounts of base are used to accelerate the reaction and to control its selectivity. Based on [{RuCl$_2$(p-cymene)}$_2$], the choice of the ligand and the added base determine the regioselectivity of the reaction.[44] If trifurylphosphine in combination with sodium carbonate is used, the Markovnikov product is formed exclusively. On the other hand, if tri-*p*-chlorophenylphosphine and 4-dimethylaminopyridine are employed, the formation of the Z-*anti*-Markovnikov product is observed (Scheme 33).

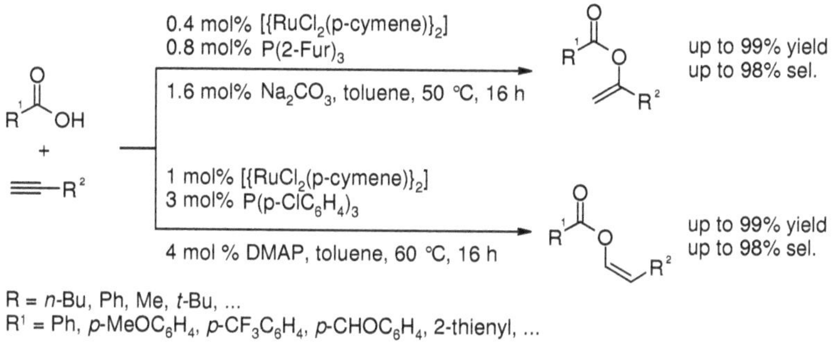

Addition of hydrogen bonded nucleophiles to alkynes

Scheme 33: addition of carboxylic acids to terminal alkynes.

The use of *N*-heterocyclic carbenes as ligands for the *anti*-Markovnikov addition described by Verpoort et al. represents an alternative to the phosphine system and also allows the control of the stereoselectivity of the ruthenium catalyzed addition of carboxylic acids to alkynes (Scheme 34).[45]

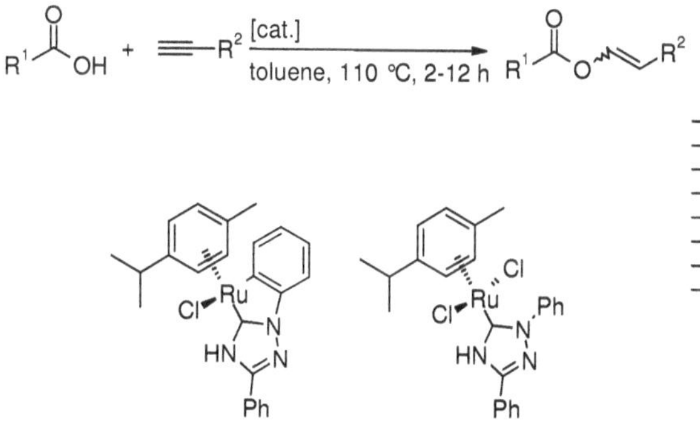

R^1	R^2	[cat.]	yield (E/Z)
Ph	H	21	88 (30/70)
Ph	H	22	100 (76/24)
Ph	CH$_3$	21	94 (22/78)
Ph	CH$_3$	22	89 (88/22)
t-Bu	CH$_3$	21	57 (91/9)
t-Bu	CH$_3$	22	93 (8/92)
n-Hex	CH$_3$	21	67 (55/45)
n-Hex	CH$_3$	22	52 (81/19)

Scheme 34: *anti*-Markovnikov addition of carboxylic acids to alkynes reported by Verpoort.

Mechanistically speaking, the formation of the Markovnikov product during the ruthenium catalyzed addition of carboxylic acids to alkynes is due to a nucleophilic attack of the carboxylic acid on a coordinated alkyne (**4**) with Markovnikov selectivity; on the other hand, Dixneuf et al. explain that the formation of the *anti*-Markovnikov product involves a vinylidene intermediate (**6**) (Scheme 35).[46]

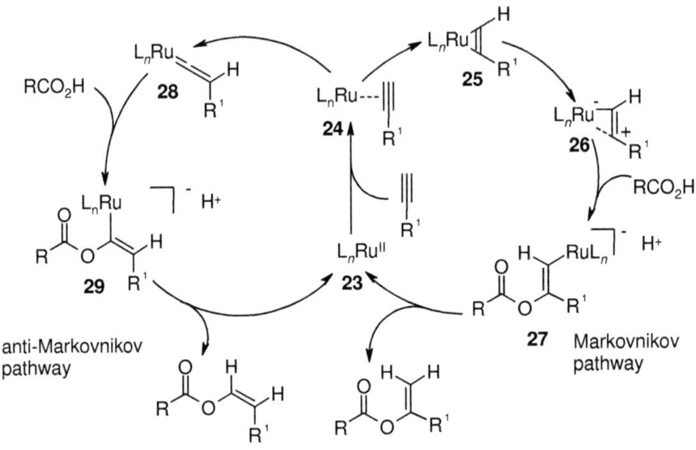

Scheme 35: proposed mechanism for the Markovnikov and *anti*-Markovnikov addition of carboxylic acids to alkynes.

After coordination of the alkyne to the ruthenium center (**24**) the intermediate could either rearrange to an alkylidene complex (**28**) or to a cyclopropene ruthenium (**25**). As vinylidenes are electron-withdrawing ligands, electron rich ligands at the ruthenium center such as electron rich phosphines or DMAP should stabilize the vinylidene intermediates formed from the alkyne and favor the nucleophilic attack on the α-position of the ruthenium center to form the intermediate **29**. After reductive elimination, the *anti*-Markovnikov product is obtained and the ruthenium(II) species **23** is released. On the other hand, if the formation of the vinylidene does not occur, formation of the intermediate **26** takes place. Since the addition of carboxylic acid is rate determining in the Markovnikov pathway, presence of a base would accelerate the formation of the vinylcomplex **27** which after reductive elimination will give rise to the Markovnikov product and regenerate the ruthenium(II) species **23**.

In addition to ruthenium, a variety of other catalytic systems have been employed for the addition of carboxylic acids to alkynes, including rhodium[47] or iridium,[48] from which the Markovnikov products are formed in good or moderate selectivities. In contrast, Hua et al. reported a system composed only from the air stable pentacarbonylrhenium(I) bromide which allows the exclusive formation of the *anti*-Markovnikov product (Scheme 36).[49] Even if the stereoselectivity outcome of the reaction is still largely dependant on the substrate, this reaction represents a non-negligible alternative to the ruthenium system and the selectivity could presumably be controlled better by the use of directing ligands.

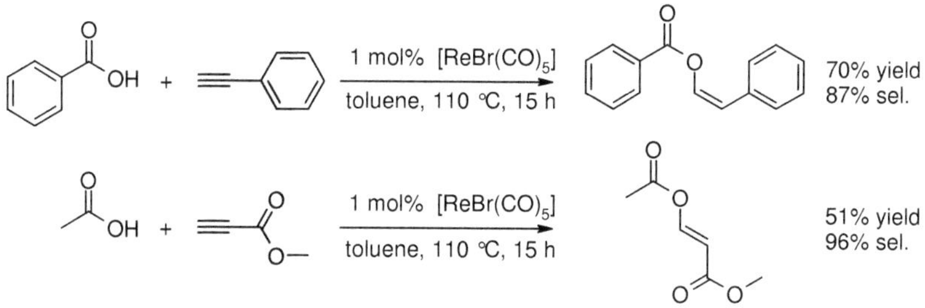

Addition of hydrogen bonded nucleophiles to alkynes

Scheme 36: Re-catalyzed addition of carboxylic acids to alkynes.

In summary, the addition of carboxylic acids to alkynes is a well established reaction. The drawback of the high toxicity bearing by mercury complexes has been overcome by using ruthenium catalysts and recent breakthroughs made the control of the regio- and stereoselectivity possible. A better understanding of the reaction mechanism could be the key for optimizing the catalyst system and increasing the selectivity and the scope of the reaction. Particularly for the *anti*-Markovnikov addition of carboxylic acids to alkynes, for which the control of the stereoselectivity is still largely dependant on the substrates used, well tuned catalytic systems or the use of new metals such as rhenium could overcome this key limitation.

1.3.3 Addition of amines to alkynes

Since ammonia and amines are much weaker acids than water or carboxylic acids, and since acids could hardly catalyze the reaction (because they would protonate NH_3 into NH_4^+), this reaction does not occur by an electrophilic mechanism and so it gives low yields unless extreme conditions are used (e.g., >200 °C).

Also, the direct addition of ammonia, primary or secondary amines to C-C multiple bonds represents the most efficient, simple and atom economic process for the formation of amines, enamines or imines (Scheme 37), which are important building blocks for organic products, e.g., pharmaceuticals, detergents, technical additives and dyes. The reaction offers an atom-efficient pathway starting from readily accessible alkenes and alkynes.

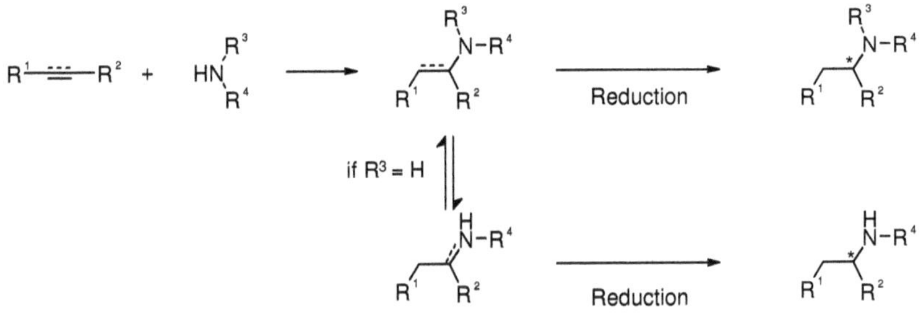

Scheme 37: addition of amines to alkenes and alkynes.

Even if the products resulting from the addition of amines to alkenes are more valuable because of their stability, the hydroamination of alkynes was for us more interesting to study because of its direct correlation with our reaction.

Thermal processes involving acetylene and ammonia usually give complex mixtures of pyridine bases. However, by controlling the speed of the gases evolved and by using supported zinc compounds as catalysts, pyridine bases are obtained only in small amounts and acetonitrile and other nitriles are formed (Scheme 38).[50]

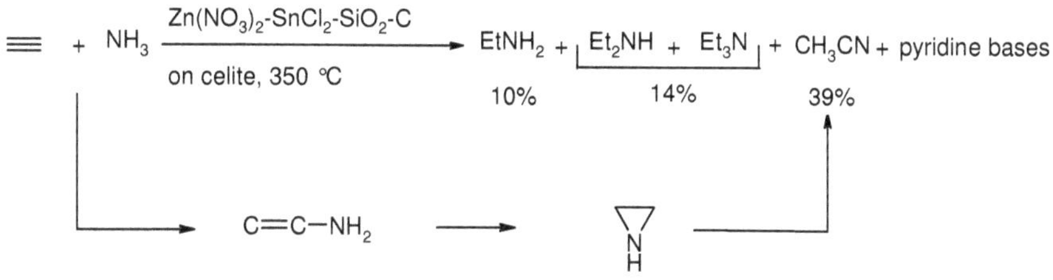

Addition of hydrogen bonded nucleophiles to alkynes

≡ + NH$_3$ —[Zn(NO$_3$)$_2$-SnCl$_2$-SiO$_2$-C on celite, 350 °C]→ EtNH$_2$ + Et$_2$NH + Et$_3$N + CH$_3$CN + pyridine bases
   10%   14%   39%

↓

C≡C–NH$_2$ → [pyrrole]

Scheme 38: reaction of acetylene and ammonia and possible intermediates.

Kozlov was probably the first who observed homogeneous catalyzed hydroamination of alkynes. During his work on the condensation of acetylene with aniline (1936) he provided experimental evidence that *N*-phenylethylideneamine is an intermediate which evolves to an aldol-type condensation product (Scheme 39).[51]

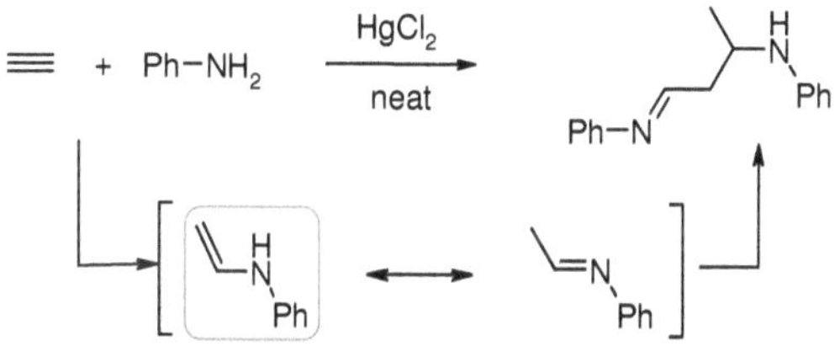

≡ + Ph–NH$_2$ —[HgCl$_2$, neat]→ Ph–N=CH–CH$_2$–NHPh

Scheme 39: hydroamination product as intermediate in the reaction of acetylene with aniline.

In 1939, based on the work of Kozlov, Loritsch et al. described the hydroamination of 1-heptyne with aniline and ethylaniline catalyzed by mercury oxide and boron trifluoride. This addition gave the corresponding enamine in low yields (7-17%).[51]

Later on (1980), Barluenga et al. found that primary and secondary aromatic amines can perform the hydroamination of terminal alkyl- and arylalkynes by using catalytic HgCl$_2$.[52] In addition, they reported that replacing the mercury catalyst by Tl(OAc)$_3$ improved the reaction if phenylacetylene is the alkyne employed (Scheme 40).[53]

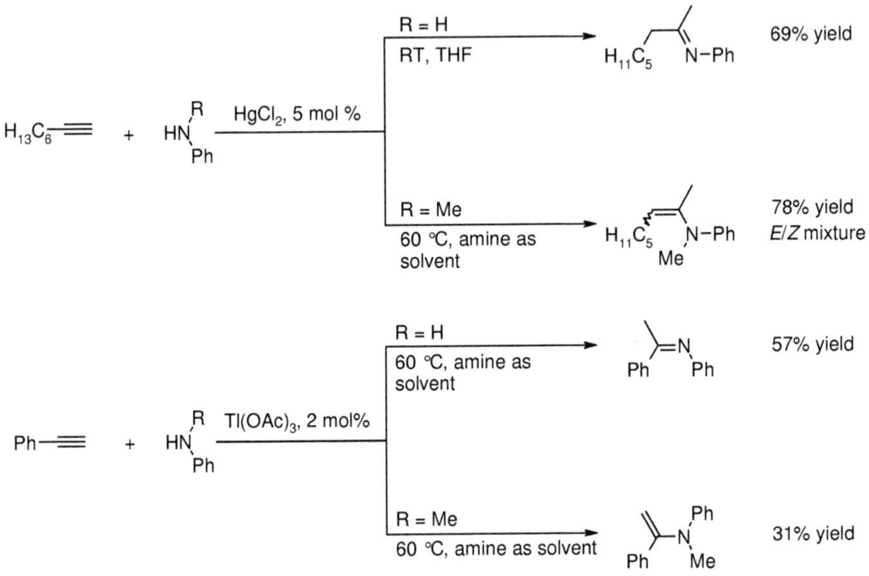

Addition of hydrogen bonded nucleophiles to alkynes

Scheme 40: Barluenga hydroamination.

Barluenga et al. presented this reaction as an aminomercuration-demercuration of alkynes which can be described as follows (Scheme 41):

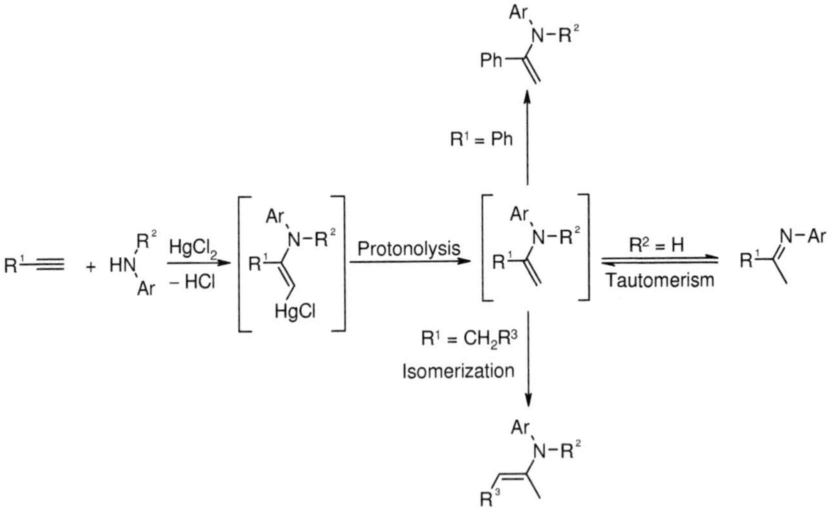

Scheme 41: aminomercuration-demercuration of terminal alkynes.

Addition of hydrogen bonded nucleophiles to alkynes

In general, the main drawback of such protocols is the high toxicity of the catalysts employed and such reactions could not be considered practical. Nevertheless, they revealed the possibility of activation of C-C triple bonds against nucleophilic attack by amines.

The developments in this area were flourishing and a plethora of methods were described during the last decades for intra- or intermolecular hydroamination of internal or terminal alkynes. An abundance of different methodologies using base[54] or metals such as lanthanides (Nd),[55] actinides (U, Th),[56] earlier transition metals (Ti, Zr)[57] and late transition metals (Au, Ru, Pd, Rh)[58] have been used to catalyze these reactions.[59] Nevertheless, there is still no general method to apply this transformation to all kinds of amines and alkynes.

Among them, catalysis based on titanium complexes has significant advantages compared to that based on some toxic (Hg, Tl) or more expensive metals (Ru, Rh, Pd, U, Th) and is often described in the literature for the hydroamination of alkynes.

Livinghouse described in 1992 an intramolecular hydroamination of amino alkynes efficiently catalyzed by $CpTiCl_3$ and $CpTi(NEt)_3Cl$ (Scheme 42, equation 1).[60] Even if these catalysts are not efficient for the intermolecular hydroamination of alkynes, several five- and six-membered cyclic imines can be synthesized. The reaction is performed under mild conditions using 20 mol% of the catalyst and the development of a process for the total synthesis of the indolizidine alkaloid (±)–monomorine with $CpTiCl_3$-catalyzed cyclization of amino alkynes as key-step demonstrates the efficiency of this transformation (Scheme 42, equation 3).[61]

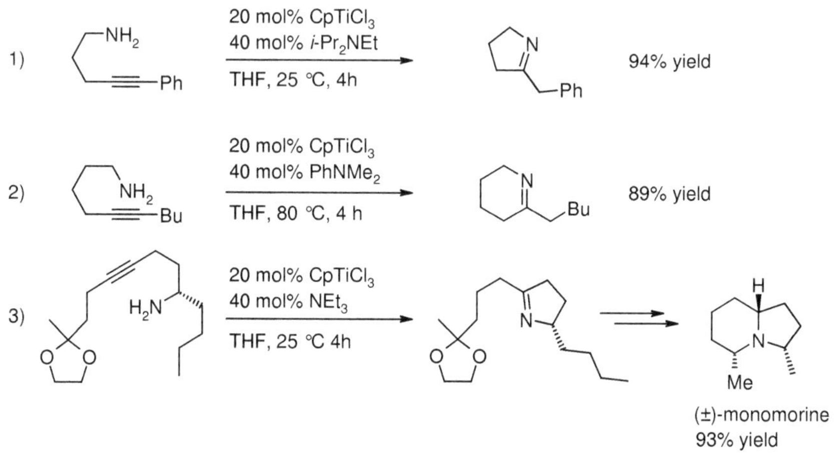

Scheme 42: intramolecular hydroamination of Livinghouse.

Addition of hydrogen bonded nucleophiles to alkynes

In 1999, based on the work of Livinghouse, Doye et al. disclosed the intermolecular hydroamination of internal alkynes. The procedure developed with Cp_2TiMe_2 as a catalyst enables primary arylamines, as well as tert- and sec-alkylamines, to be coupled with several disubstituted alkynes in good yields,[62] although primary n-alkylamines and terminal alkynes can only be coupled in poor yields. In general, the more favored product bears the smaller alkyne substituent located α to the nitrogen atom. The initially formed imines are usually hydrolyzed to ketones or reduced to secondary amines (Scheme 43).

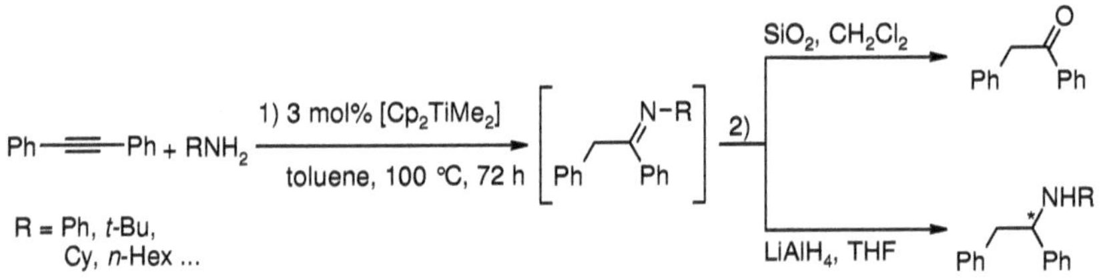

Scheme 43: Doyle's intermolecular hydroamination.

An interesting application for this reaction is the hydroamination of internal alkynes using α–aminodiphenylmethane as an ammonia equivalent. This transformation, even if its not atom economic in view of the amounts of waste produced during the process, represents an alternative to the tricky direct addition of ammonia to the triple bond and allows the synthesis of primary amines. The primary imines formed can be hydrogenated and cleaved directly to the corresponding primary amines by catalytic hydrogenation using Pd/C as catalyst (Scheme 44).[63]

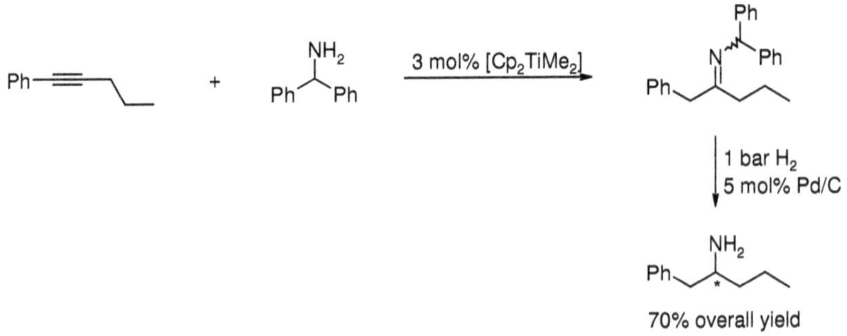

Scheme 44: ammonia equivalent hydroamination of internal alkynes.

Addition of hydrogen bonded nucleophiles to alkynes

Nevertheless, the main drawbacks of this transformation are the elevated temperatures (100-120 °C) and the long reaction times (up to 72 h) that are usually required for the reaction to proceed. This issue could be overcome by using microwaves which reduces the reaction time by a factor of 10. Surprisingly, the use of microwave irradiation enhanced the poor yielding hydroamination of terminal alkynes, bringing the transformation to almost full conversion (Scheme 45). Moreover, these new conditions enabled the first *anti*-Markovnikov hydroamination of alkynes. Indeed, while the use of long-chain alkynes such as 1-dodecyne yields predominantly a Markovnikov product, the use of phenylacetylene leads to the formation of the *anti*-Markovnikov product. The responsible factors for this switch in regioselectivity are not yet determined.[64]

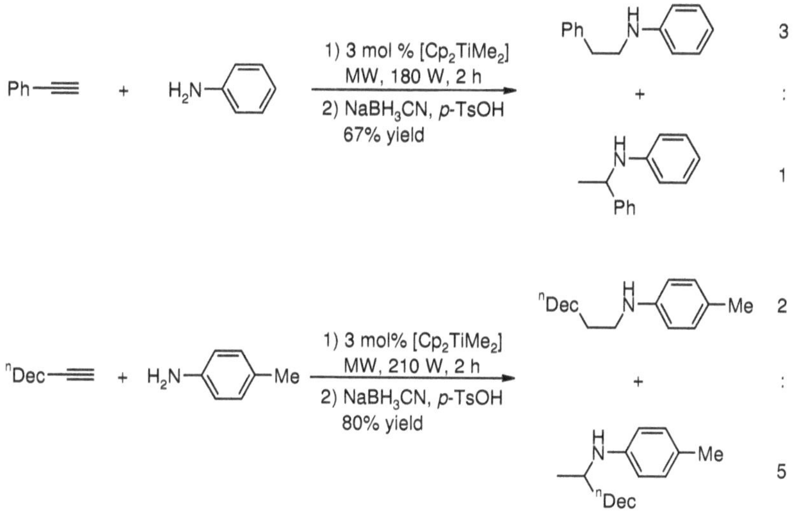

Scheme 45: microwave assisted hydroamination of terminal alkynes.

From a mechanistic point of view, Bergman's investigations show that the catalytically active species of the described reaction is a cyclopentadienyl(amido)titanium complex.[65] Kinetic investigation performed by Doyle et al. in combination with the mechanistic studies of Bergman allowed them to suggest a catalytic cycle which correctly described the hydroamination of alkynes (Scheme 46).

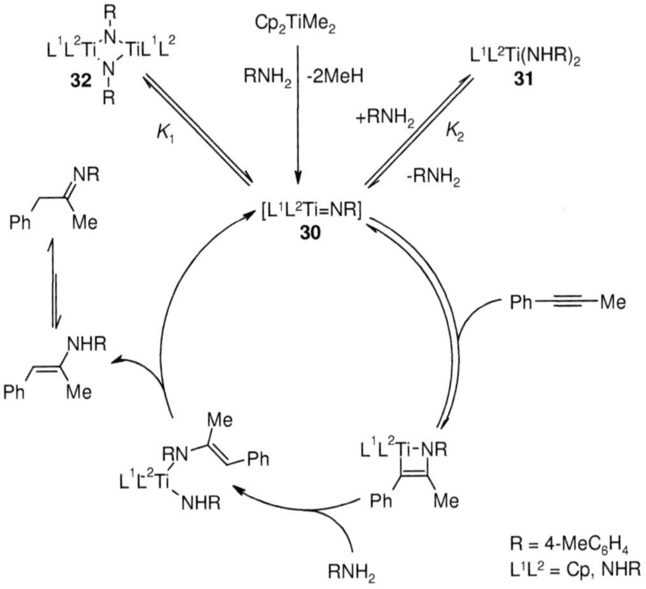

Scheme 46: catalytic cycle for the titanium catalyzed hydroamination of alkynes.

The unfavorable equilibriums (K_1, K_2) between the imido complexes (**29**), the imido complex dimers (**32**) and bisamides (**31**) explain the fact that sterically demanding amines are better substrates than the less hindered amines for the Cp_2TiMe_2-catalyzed hydroamination of alkynes. Kinetic studies suggest that bigger ligands at the titanium center should influence these equilibriums and favor the reaction of less hindered amines. The proof of concept was made by using $Cp^*_2TiMe_2$ which catalyzed the reaction between *n*-propylamine and diphenylacetylene leading after subsequent reduction to the corresponding secondary amine in 86% yield, while the use of Cp_2TiMe_2 for the same reaction only gives 10% of the desired product (Scheme 47).[66]

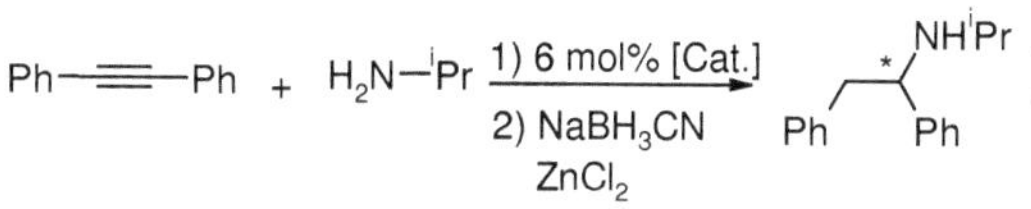

Ph─≡─Ph + H$_2$N─^{i}Pr $\xrightarrow[\substack{2)\ NaBH_3CN \\ ZnCl_2}]{1)\ 6\ mol\%\ [Cat.]}$ Ph⧸*⧹Ph (NH^{i}Pr)

catalyst	reaction time	yield
Cp*$_2$TiMe$_2$	4 h	86%
Cp$_2$TiMe$_2$	48 h	10%

Scheme 47: comparison between Cp$_2$TiMe$_2$ and Cp*$_2$TiMe$_2$ for the hydroamination of alkynes by *n*-propylamine.

On the other hand, the use of late transition metals for the hydroamination represents a great advantage because of their lower affinity to oxygen. Ruthenium catalysts seem to represent the best compromise in terms of functional group compatibility, toxicity and costs.

In 1999, Uchimaru et al. disclosed the first ruthenium-catalyzed hydroamination of alkynes. In the presence of Ru$_3$(CO)$_{12}$, phenylacetylene and its derivatives undergo the regioselective Markovnikov insertion of N-H bond of *N*-methylaniline to afford *N*-methyl-*N*-(α-styryl)anilines in good yields (Scheme 48).[67]

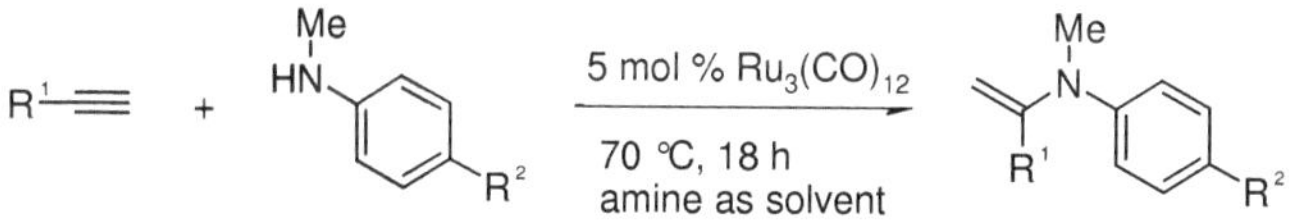

R^1─≡ + HN(Me)─C$_6$H$_4$─R^2 $\xrightarrow[\substack{70\ °C,\ 18\ h \\ amine\ as\ solvent}]{5\ mol\ \%\ Ru_3(CO)_{12}}$ R^1─C(=CH$_2$)─N(Me)─C$_6$H$_4$─R^2

R^1	R^2	yield
Ph	H	85%
p-MeC$_6$H$_4$	H	78%
p-FC$_6$H$_4$	H	88%
Ph	Me	76%
Cyclohex-1-enyl	H	77%

Scheme 48: ruthenium catalyzed hydroamination of alkynes.

A plausible mechanism for this transformation is the one proposed by Uchimaru, based on the (amido)ruthenium hydride as intermediate (**34**) (Scheme 49).

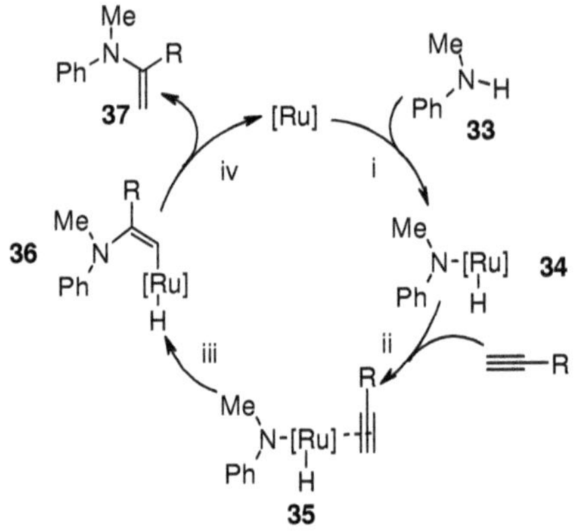

Addition of hydrogen bonded nucleophiles to alkynes

Scheme 49: plausible mechanism for the ruthenium catalyzed hydroamination of alkynes.

This mechanism involves (i) oxidative addition of the N-H bond of amine **33** to the ruthenium center giving the (amido)ruthenium hydride **34**. Coordination of the alkyne to the ruthenium center leads to the complex **35**, which undergoes insertion of the coordinated carbon-carbon triple bond into the Ru-N bond. Finally, the reductive elimination of the enamine **37** from the vinyl ruthenium species **36** regenerates the coordinatively unsaturated ruthenium(0) center. It is worth mentioning that the reaction proceeds regioselectively to give the Markovnikov addition product. The only drawback of this process is the necessity to use a ten-fold excess of the amine derivative to ensure high yields. Using only a five-fold excess of the amine resulted in a dramatic reduction of the yields, typically lower than 26%. Isomers of the enamine and trimers of the employed phenylacetylenes are detected as side products.

A ruthenium/acid catalyst system was introduced nearly at the same time by Wakatsuki et al.[68] allowing the reaction of aniline derivatives with terminal alkynes with formation of the corresponding imines. The use of the strong acids HPF_6 and HBF_4 and their ammonium salts as additives dramatically enhances the rate and the yield of the reaction. For example, the reaction between aniline and phenylacetylene was completed in 12 h at 100 °C with only 0.1 mol% $[Ru_3(CO)_{12}]$ and 0.3 mol % NH_4PF_6 and gave a 84% yield (GC) of the corresponding product, while only a 2.6% yield of the product was obtained under the same conditions but without the additive (Scheme 50). The effect of the acid remains to be clarified, but it has been reported by Lavigne et al. that halides and related salts can promote substitution reactions of $[Ru_3(CO)_{12}]$.[69]

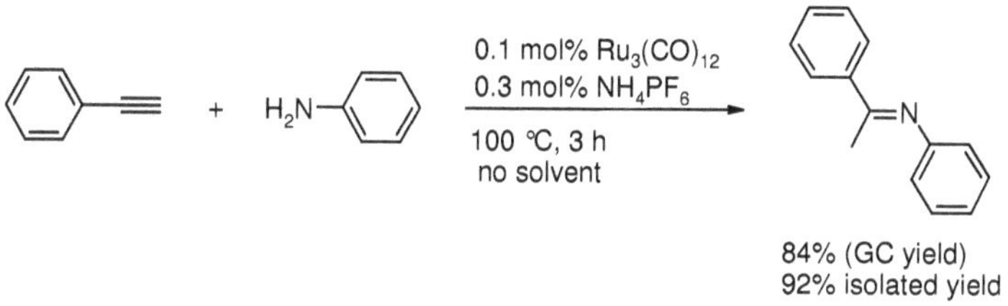

Scheme 50: addition of aniline to phenylacetylene.

Based on this reaction Wakatsuki and Tokunaga et al. developed a new Bischler-type indole synthesis.[70] This ruthenium-catalyzed one-pot reaction allows the synthesis of 2-substituated-3-methyl indols in good yields and selectivities starting from commercially available propargylic alcohols and aniline (Scheme 51).

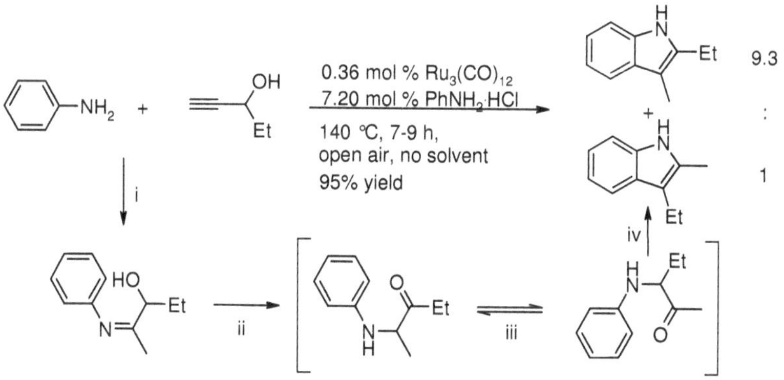

Scheme 51: Bischler-type indole synthesis by Wakatsuki and Tokunaga.

The reaction sequence consists of three steps: hydroamination of C-C triple bond of the propargylic alcohol (i), hydrogen migration of the resulting iminoalcohol to the aminoketone (ii) and cyclization giving rise to the indole scaffold (iv). Moreover, it is known that there is apparently a fast interconversion of the regioisomers of aminoketones in the presence of aniline hydrochloride so that different aminoketones often give the same indole (iii).[71]

Finally, one example for the ruthenium-catalyzed *anti*-Markovnikov hydroamination of alkynes is the reaction of 1,1-dimethylhydrazines with terminal alkynes. This transformation developed by Fukumoto et al. is catalyzed by TpRuCl(PPh$_3$)$_2$ and yields nitriles as products.[72] Mechanistically speaking, a vinylidene species is proposed as the key intermediate for this reaction. The addition of the hydrazine to the vinylidene ligand generates a hydrazine carbene, which loses dimethylamine and provides the nitrile (Scheme 52).

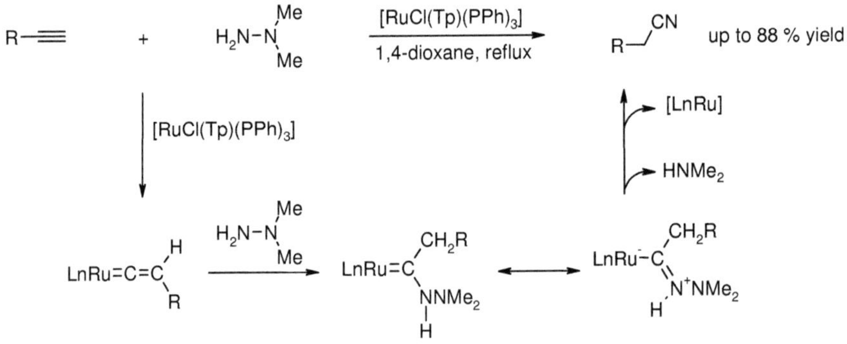

Scheme 52: ruthenium catalyzed addition of dimethylhydrazine to alkynes.

In summary, great success has been made for the hydroamination of alkynes since Kozlov's work. Titanium complexes bearing two labile ligands seem to be the catalysts of choice for this reaction, but the functional group tolerance is supposed to be low because of the high oxophilicity of this metal.

Late transition metal based complexes, and particularly ruthenium catalysts, offer theoretically a better group tolerance. Nevertheless, the scope of such processes is still often limited to specific substrates. However, this chemistry is under intensive studies and there is no doubt that this drawback will be overcome thus making late transition metals as important as titanium for the hydroamination of alkynes.

1.3.4 Addition of amides to alkynes

As shown before, the enamide group is an important substructure. Nevertheless in comparison with enamine's chemistry, the chemistry of enamides has not been well developed, and synthetic methods for them are rather limited. Although, the product resulting from the addition of amides to alkynes are much stable and often lead to more important products. A catalytic addition of amides to alkynes (or hydroamidation of alkynes) would be an ideal synthetic entry to enamides, since it would use readily available starting materials and be inherently atom-economic.

Similarly to addition of water, carboxylic acids and amines to alkynes, the amides can add to alkynes regioselectively to provide Markovnikov and *anti*-Markovnikov products (Scheme 53).

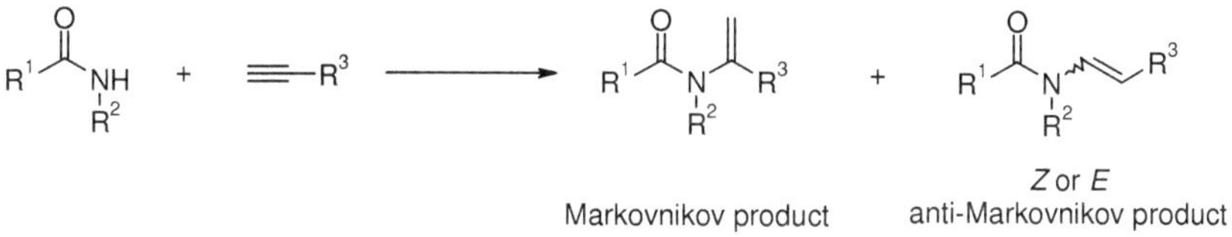

Scheme 53: addition of amides to alkynes.

This chemistry does not have a long history like the addition of water to alkynes, and no catalysis based on mercury has been developed. The first example of transition metal complex-catalyzed addition of amides to alkynes was described in 1995. Based on previous work on the addition of carboxylic acids to alkynes, Watanabe et al. developed a system capable of catalyzing the addition of amides to alkynes.[73] The authors chose a system based on the low valent ruthenium dodecacarbonyl complex [$Ru_3(CO)_{12}$] to perform the reaction. The catalyst system formed *in situ* from [$Ru_3(CO)_{12}$] associated to tricyclohexylphosphine (PCy$_3$) enables the direct *anti*-Markovnikov addition of *N*-aryl substituted amides to terminal alkynes and provides the corresponding *E*-isomers predominantly in fair yields (Scheme 54).

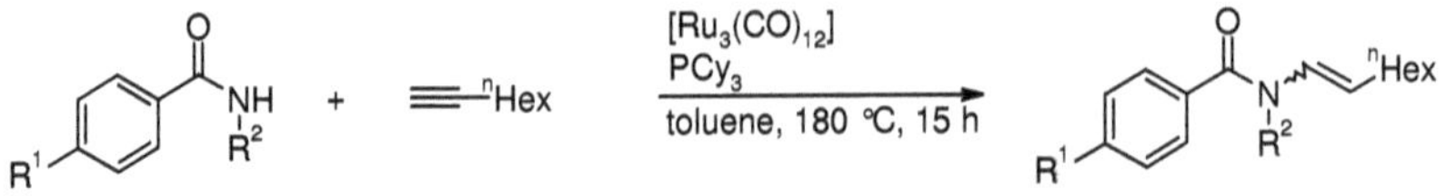

Addition of hydrogen bonded nucleophiles to alkynes

R^1	R^2	yield (E/Z)
H	H	67 % (93/7)
OMe	H	58 % (89/11)
Cl	H	49 % (85/15)
H	Me	65 % (100/0)

Scheme 54: ruthenium catalyzed addition of amides to alkynes.

Insertion of a triple bond into a Ru-H bond resulting from a N-H activation was proposed as the preferred mechanism (Scheme 55).

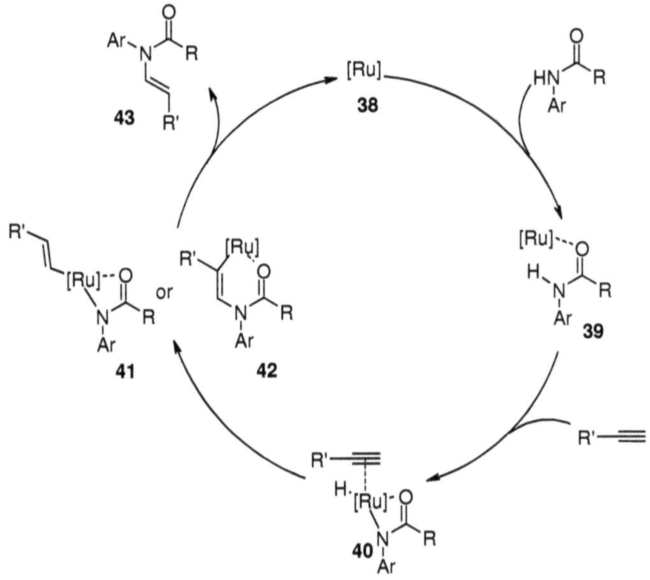

Scheme 55: proposed mechanism for the hydroamidation of alkynes from Watanabe.

The authors explain the regioselectivity of the reaction by proposing the complex **39** resulting from the amide coordination to the ruthenium complex through its carbonyl oxygen atom as a key intermediate. This complex leads to an intermediate **40** in which the N-H bond would be activated. Then the alkyne coordination and the concomitant oxidative addition of the N-H bond followed by an alkyne insertion gives an intermediate **41** or **42**. A subsequent reductive elimination gives the enamide **43**, regenerating the ruthenium species.

Nevertheless, only a very limited range of *N*-aryl substituted amides can be added to 1-octyne using this transformation, albeit at extremely high temperatures (180 °C). Clearly, much more effective catalyst systems are required to allow an application of this reaction in organic synthesis.

A major breakthrough was made recently in this area by Gooßen et al.[74] Indeed, they finally overcame the problem of the low scope by using another source of ruthenium. With a system based on the commercially available ruthenium dimethallyl cyclooctadiene [Ru(cod)(met)$_2$], tributylphosphine (*n*-Bu$_3$P) and DMAP, *E*-enamides were formed regio- and stereoselectively (Scheme 56).

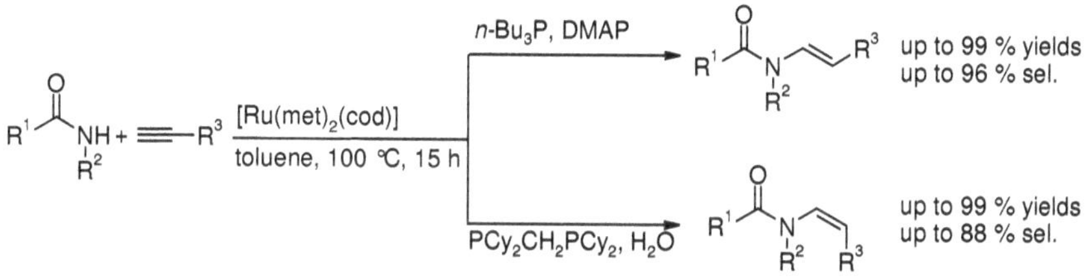

Scheme 56: hydroamidation of alkynes from Gooßen et al.

Compared to the method of Watanabe, numerous alkynes were successfully converted into enamides using this protocol. Among them phenylacetylene, acetylenecarboxylic esters, trimethylsilylacetylene, and even conjugated enynes. Besides lactams, various secondary amides, anilides, ureas, bislactams, carbamates can be used as the N-H-coupling partner. Under similar conditions, but in the presence of bisdicyclohexylphosphinomethane (Cy$_2$PCH$_2$PCy$_2$) and water as additive, the catalytic *anti*-Markovnikov addition selectively provides the Z isomers. It is noteworthy that this reaction allows the use of much lower temperatures than the previous, better results in yields and selectivity were obtained by only heating the reaction at 100 °C.

An alternative mechanism to the one proposed by Watanabe was postulated by Dixneuf et al. for this transformation. In his review article about metal vinylidenes in catalysis, he applied the same mechanism as for the addition of carboxylic acids to alkynes to this reaction. This way he explained that this transformation probably involves formation of an alkylidene intermediate (Scheme 57).

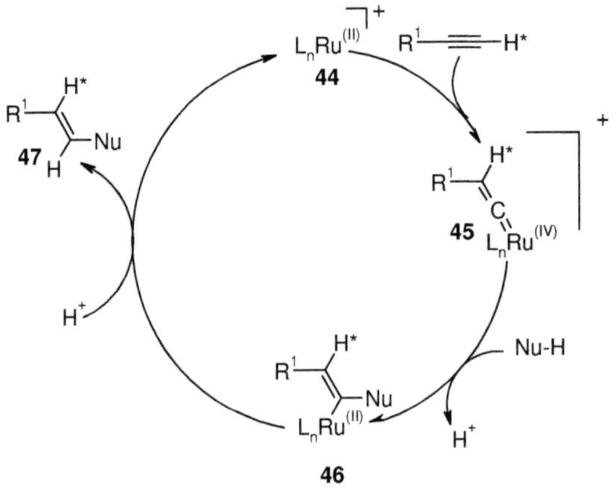

Scheme 57: Dixneuf's proposed mechanism for the hydroamidation of alkynes.

An activated cationic ruthenium **44** species binds to the alkyne through a 1,2-proton shift to form an alkylidene intermediate **45**. A nucleophilic attack in α-position of the ruthenium center of the alkylidene intermediate forms the intermediate **46**. Protonolysis of the latter releases the enamide **47** and regenerates the ruthenium active species.

Nevertheless, this mechanism is still unclear and intensive studies are required to understand the multiple steps of this catalytic process. Moreover, at this point the reaction is still exclusively limited to secondary amides. Other classes of amides such as primary amides, imides or thioamides are inactive with the proposed system or were not yet tested. For primary amide substrates, which would give access to the synthetically far more valuable secondary enamides, either no conversion at all was observed, or mostly double vinylation products in traces and as mixtures of (E/Z)-isomers were reported.

On the other hand, protocols for the Markovnikov hydroamidation of alkynes are barely found in the literature. An interesting method for the synthesis of dehydroamino acids has been described by Trost in 1997,[75] who added N–H nucleophiles to alkynoates under formation of 2-N–H substituted acrylates. Phosphine as a catalyst induces isomerization of the alkynoates to 2,4-dienoates, which then undergo addition of pronucleophiles to the 4-position (Scheme

Addition of hydrogen bonded nucleophiles to alkynes

58). Trost optimized a system capable of adding imides to some terminal alkynes to give the Markovnikov product, by only using triphenylphosphine, sodium acetate and acetic acid.

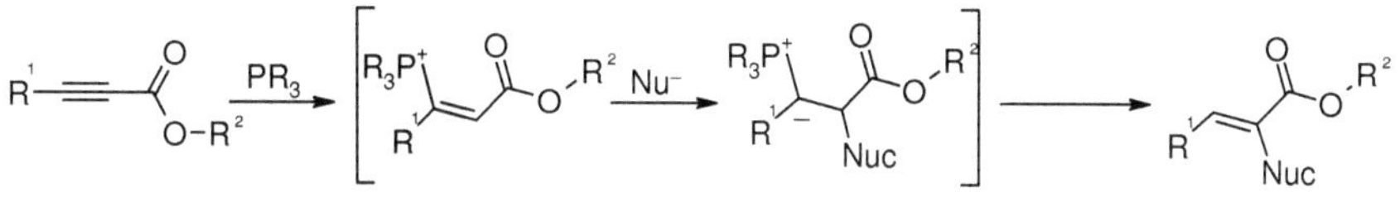

Scheme 58: Alkynoates-nucleophilic additions at the α-position.

Unfortunately, only specific classes of substrates could be converted by using this reaction. Only succinimides and phtalimides add to Michael acceptor type terminal alkynes such as methyl- or ethylpropiolate, and progress also needs to be made in this area.

1.4 Project aim

A thorough survey of the specialty literature provided represented a great source of inspiration for our work and an efficient catalyst was developed in our group for the addition of secondary amides to alkynes. However, this transformation is strictly limited to this specific class of substrates and only the product resulting from an *anti*-Markovnikov addition could be obtained. Great applications could be envisioned, particularly in natural product syntheses, if we overcome these key limitations.

The aims of the present study were:

- To develop a new catalytic system for the addition of new classes of amides to alkynes:
    - Since we already observed activity in the case of the imides, the optimization of the reaction for this substrate class was the starting point of our studies.
    - Particularly interesting substrate candidates are the primary amides. However, it is possible to envisage the complexity of this reaction where we had to overcome the low nucleophilicity of the primary amides, while avoiding the addition of the monovinylated product to a second equivalent of alkynes.
    - Finally we concentrated our efforts in finding a catalytic system active for the addition of thioamides to alkynes. Thioamides have a more ambident nucleophilic character than amides do, and thus can react at the sulfur terminus rather than the nitrogen one. Therefore, it will be crucial to have a perfect control on the chemo-, regio- and stereoselectivity of the reaction.
- To have a better understanding of the reaction. Thus mechanistic studies were performed by using *in situ* NMR and ESI-MS in order to identify the active species formed during the reaction.
- To find a new catalytic system based on a less expensive ruthenium sources in order to replace the [Ru(cod)(met)$_2$] used in our previous protocol for the addition of secondary amides to alkynes.
- To reverse the regioselectivity of the reaction in favor of the Markovnikov product.

Ruthenium is a very low oxophilic metal and displays the widest scope of oxidation states (from -2 valent in Ru(CO)$_4^{2-}$ to octavalent in RuO$_4$) and therefore has great potential for the exploitation of novel catalytic reactions and synthetic methods. Ruthenium also presents very low toxicity compared to mercury-based catalysts; actually, its potential for the medical field is now being recognized (ruthenium anticancer agents).

As shown previously ruthenium based catalysts could act either as Lewis acids and thus promote Markovnikov addition of hydrogen-bonded nucleophiles to alkynes or be involved in vinylidene intermediates and lead to *anti*-Markovnikov products.

This important feature in addition to the low prices of ruthenium complexes compared to other metals made us choose ruthenium complexes for our further investigations.

2 Results and discussion

2.1 Addition of imides to alkynes

2.1.1 Strategy for the development of catalyst system for the addition of imides to alkynes

The catalyst system previously developed for the addition of secondary amides to alkynes is not extendable to the imides substrates. Different reasons could explain this lack of reactivity; among them the acidity of the imide proton (situated on the nitrogen) which is much more acidic than in the case of the secondary amides (Table 1) and thus makes imides less nucleophilic than secondary amides.

Table 1: pK_a of some compounds in DMSO.[76]

compound	pK_a
benzoic acid	11.1
acetic acid	12.6
succinimide	14.7
1-methylhydantoin	15.0
pyrrolidinone	24.2
2-piperidone	26.6

The aim of this project is to find a catalyst system able to promote the addition of a wide range of imides to alkynes. This well tuned catalyst system, while showing high activity with regard to the imides, should also be inherently atom-economic and easy to handle.

As a starting point for catalyst development, we used the reaction of succinimide (**1a**) with 1-hexyne (**2a**) to investigate the catalytic activity of several ruthenium complexes under various conditions (Scheme 59).

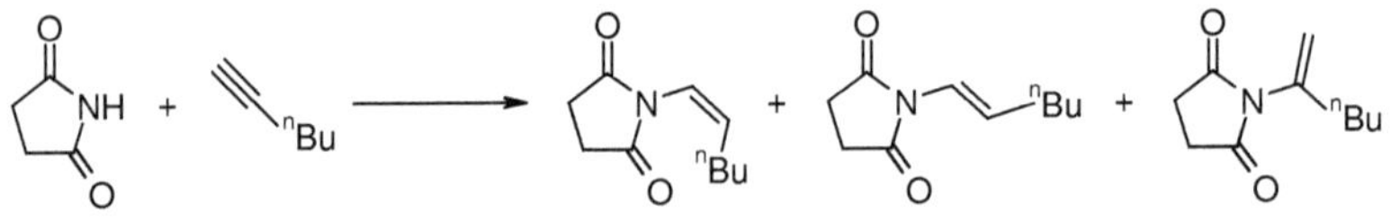

Scheme 59: addition of succinimide to 1-hexyne.

Addition of imides to alkynes

In theory, three different products could be obtained by addition of succinimide (**1a**) to alkynes (**2a**): two products resulting from an *anti*-Markovnikov addition (**3aa** and **4aa**) and one resulting from a Markovnikov addition (**5aa**). Nevertheless, only the *anti*-Markovnikov products were observed during this project. This project was first carried out in collaboration and then taken over from Dipl.-Chem. Claus Brinkmann. He had developed the first effective catalyst within his diploma work and concluded from spectroscopic data that the products that he isolated had *E*-configuration. After successful isolation of both isomers, I performed additional NMR experiments, which led to the conclusion that the products isolated by Claus Brinkmann were in fact the Z-isomers. I thus repeated the decisive screening experiment and isolated the products once again to confirm their stereochemistry as detailed below.

2.1.2 Development of a new catalyst system

The first test we did by using the exact same system as for the addition of secondary amides to alkynes developed by Gooßen et al. showed, as expected, very low activity (Scheme 60).

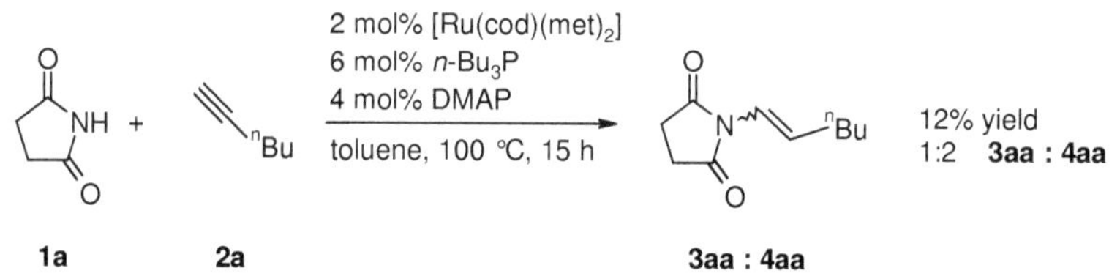

$$\text{1a} + \text{2a} \xrightarrow[\text{toluene, 100 °C, 15 h}]{\substack{2 \text{ mol\% [Ru(cod)(met)}_2] \\ 6 \text{ mol\% } n\text{-Bu}_3\text{P} \\ 4 \text{ mol\% DMAP}}} \text{3aa : 4aa} \quad \begin{array}{l} 12\% \text{ yield} \\ 1{:}2 \quad \textbf{3aa : 4aa} \end{array}$$

Scheme 60: addition of succinimide to 1-hexyne with the conditions of the addition of secondary amides to alkynes.

Nevertheless, we made a critical observation during the reaction concerning the low solubility of succinimide in toluene. We first tested the reaction in different solvents and found out that dimethylformamide (DMF), while preserving the yields and ratio of the reaction, was also able to dissolve the starting materials. We then decided to opt for DMF as a solvent for the further development of the reaction.

Variation of the ruthenium sources

We chose ruthenium as a metal center for the development of a new catalytic system, because of its high activity in alkyne chemistry and particularly in the addition of amides to alkynes. For the ligand, tri-n-butylphosphine (n-Bu$_3$P), which was decisive for the addition of secondary amides to alkynes, was also used as a starting point for the development of the new catalytic system.

Different ruthenium sources were then tested in DMF as depicted in the table below.

Table 2: test of different ruthenium sources.

entry	[Ru-cat.]	yield (%)[a]	ratio (3aa:4aa)[b]
1	[Ru(cod)(met)$_2$]	8	2:1
2	[Ru$_3$(CO)$_{12}$]	0	-
3	[Ru(acac)$_3$]	0	-
4	[RuCl$_2$(CO)$_3$]$_2$	0	-
5	[RuCl$_2$(C$_6$H$_6$)]$_2$	13	1:1
6	[(p-cymene)RuCl$_2$]$_2$	15	1:1

Reaction conditions: 1 mmol succinimide, 2 mmol 1-hexyne, 2 mol% Ru-source, 6 mol% n-Bu$_3$P, DMF, 100 °C, 15 h. a) corrected GC yields with n-tetradecane as internal standard. b) ratio determined by GC.

As shown in the table above, [Ru(cod)(met)$_2$] (entry 1) and [(p-cymene)RuCl$_2$]$_2$ (entry 6) previously used for the addition of secondary amides to alkynes and the addition of carboxylic acids to alkynes, respectively, showed promising results. While [Ru(met)$_2$(cod)] already showed better selectivity, [(p-cymene)RuCl$_2$]$_2$ was able to promote the reaction in a better yield. Since the selectivity of the catalyst is as much important as its activity, we decided to run the next set of experiments using in parallel [Ru(met)$_2$(cod)] and [(p-cymene)RuCl$_2$]$_2$.

The effect of the additives

As experienced during the previous work dealing with the addition of secondary amides to alkynes, the presence of an additive is crucial. We therefore studied the effect of different additives (e.g. bases, acids, salts) on the reaction.

Table 3: test of different additives.

entry	[Ru-cat.]	additive	yield (%)[a]	ratio (3aa:4aa)[b]
7	[Ru(cod)(met)$_2$]	CH$_3$COOH	0	-
8	[(p-cymene)RuCl$_2$]$_2$	"	0	-
9	[Ru(cod)(met)$_2$]	DMAP	18	2:1
10	[(p-cymene)RuCl$_2$]$_2$	"	14	1:1
11	[Ru(cod)(met)$_2$]	NEt$_3$	16	1:1
12	[(p-cymene)RuCl$_2$]$_2$	"	12	1:1
13	[Ru(cod)(met)$_2$]	MgCO$_3$	20	2:1
14	[(p-cymene)RuCl$_2$]$_2$	"	19	1:1
15	[Ru(cod)(met)$_2$]	KBr	12	2:1
16	[(p-cymene)RuCl$_2$]$_2$	"	14	1:1
17	[Ru(cod)(met)$_2$]	Mg(OTf)$_2$	15	3:1
18	[(p-cymene)RuCl$_2$]$_2$	"	13	1:1

Reaction conditions: 1 mmol succinimide, 2 mmol 1-hexyne, 2 mol% Ru-source, 6 mol% n-Bu$_3$P, 4 mol% additive, DMF, 100 °C, 15 h. a) corrected GC yields with n-tetradecane as internal standard. b) ratio determined by GC.

As expected, the use of acid totally stops the reaction (entries 7, 8). Conversely, the use of base (e.g. DMAP) enhances the rate of the reaction, particularly if [Ru(cod)(met)$_2$] is employed (entry 9). An interesting result was obtained by using MgCO$_3$; in combination with [Ru(cod)(met)$_2$] (entry 13), a significant 20% yield was obtained while retaining the selectivity previously observed. Interestingly, the use of Mg(OTf)$_2$ shows better selectivity (entry 17). Since both catalysts showed almost the same activity, but with a significant influence on the selectivity if [Ru(met)$_2$(cod)] is employed, we decided to only employ the latter for further optimizations.

Suspecting that the Lewis acidic magnesium ion might have a share in the beneficial effect of this additive, we extended our search to other Lewis acids (Table 4).

Table 4: test of different Lewis acids.

entry	additive	yield (%)[a]	ratio (3aa:4aa)[b]
17	Mg(OTf)$_2$	15	3:1
19	Zn(OTf)$_2$	44	6.1
20	Al(OTf)$_3$	59	6:1
21	Yb(OTf)$_3$	98	7:1
22	Sc(OTf)$_3$	90	6:1
23	Yb(NO$_3$)$_3$	80	6:1

Reaction conditions: 1 mmol succinimide, 2 mmol 1-hexyne, 2 mol% Ru-source, 6 mol% n-Bu$_3$P, 4 mol% additive, DMF, 100 °C, 15 h. a) corrected GC yields with n-tetradecane as internal standard. b) ratio determined by GC.

We were pleased to find that they caused a dramatic increase in catalyst productivity. Scandium and ytterbium triflates were particularly effective, leading to complete conversion of the starting imide and the formation of the desired product in good selectivities (entries 21, 22). This is probably due to a coordination of the Lewis acid to the oxygen atom of the amide thus enhancing its nucleophilicity.

The effect of the ligands

Since the choice of the ligand could have significant influence on the selectivity of a metal-catalyzed reaction, we next set up a range of experiments to determine if it is possible to either increase the selectivity for the Z-enimide or totally reverse the selectivity in favor of the E-enimide.

Table 5: test of different ligands.

entry	ligand	yield (%)[a]	ratio (3aa:4aa)[b]
21	n-Bu$_3$P	98	7:1
22	P(2-pyridyl)Ph$_2$	19	1:1
23	PCy$_2$(CH)$_2$PCy$_2$	61	8:1
24	P(2-furyl)$_3$	4	2:1
25	1,1'-bis-(PPh$_2$)-ferrocene	0	-
26	Imes(n-Bu)$_3^+$Cl$^-$ [c]	0	-
27	PPh$_3$	53	1:2
28	i-Pr$_3$P	35	1:15
29	PCy$_3$	24	1:5

Reaction conditions: 1 mmol succinimide, 2 mmol 1-hexyne, 2 mol% [Ru(cod)(met)$_2$], 6 mol% ligand, 4 mol% Yb(OTf)$_3$, DMF, 100 °C, 15 h. a) corrected GC yields with n-tetradecane as internal standard. b) ratio determined by GC. c) 1,3-dibutylimidazolium chloride.

Unfortunately, no increase of the Z-selectivity was observed and n-Bu$_3$P remains the best ligand for this protocol (entries 21-26). However, we discovered that with tri-isopropylphosphine (i-Pr$_3$P), the E-isomer became the major product (entry 28). This change in selectivity was also observed if triphenylphosphine or tricyclohexylphosphine are employed (entries 27 and 29, respectively) and it is probably due to the steric hindrance of these ligands.

Optimization of the reaction conditions

Finally a range of experiments was set up to find out the ideal conditions for this reaction. It was important to determine whether or not humidity or air can influence the activity of the catalyst, if a less toxic solvent than DMF could be used and if the temperature and the time of the reaction could be reduced.

Table 6: optimization of the reaction conditions.

entry	solvent	temperature (°C)	yield (%)[a]	ratio (3aa:4aa)[b]
21	DMF	100	98	7:1
30	toluene	"	5	2:1
31	acetonitrile	"	0	-
32	THF	"	0	-
33	NMP	"	25	4:1
34[c]	DMF	"	96	6:1
35[d]	DMF	"	0	-
36	"	80	98	7:1
37	"	60	98	7:1
38	"	25	45	10:1
39[e]	"	60	98	7:1

Reaction conditions: 1 mmol succinimide, 2 mmol 1-hexyne, 2 mol% [Ru(cod)(met)$_2$], 6 mol% n-Bu$_3$P, 4 mol% Yb(OTf)$_3$, 15 h. a) corrected GC yields with n-tetradecane as internal standard. b) ratio determined by GC. c) 3 mmol of water are added. d) air was introduced into the vessel during the reaction. e) the reaction was stopped after 6 h.

In comparison to DMF, other solvents were less effective: low polarity solvents were unable to dissolve the polar reagents and catalyst, while sufficiently strong coordinating solvents, e.g. acetonitrile, appeared to slow down the reaction (entries 30-33). The addition of water does not affect the reaction (entry 34) but if air is introduced in the vessel, the reaction stops immediately (entry 35), probably because of the high sensitivity of tri-n-butylphosphine. A reaction temperature of 60 °C turned out to be sufficient (entry 37), and a further decrease to

Addition of imides to alkynes

25 °C led to an increase in Z-selectivity, though with a lower yield (entry 38). Finally after only 6 h, the reaction appeared to be complete (entry 39).

2.1.3 Scope of the reaction

After optimizing the catalyst system, we tested the generality of the reaction protocols with regard to both coupling partners. An excess of the alkyne was used to make up for losses due to evaporation or oligomerization reactions. Since scandium(III)trifluoromethanesulfonate (Sc(OTf)$_3$) shows almost the same activity as ytterbium(III)trifluoromethanesulfonate (Yb(OTf)$_3$) while having a higher tolerance in regard to functional groups, we preferred to choose Sc(OTf)$_3$ to perform the reactions.

As shown in table below, several imides were added smoothly to 1-hexyne in high yields and selectivities.

Table 7: addition of different imides to 1-hexyne.

imide	product	yield (%)[a]	ratio (3:4)[b]
1a	3aa	98	8:1
1b	3ba	58 (72)[c]	14:1
1c	3ca	62 (81)[c]	9:1

- 55 -

Addition of imides to alkynes

imide	product	yield (%)[a]	ratio (3:4)[b]
1d (phthalimide)	3da	88	4:1
1e (pyridine-fused)	3ea	14	3:1
1f (diacetamide)	3fa	8[d]	1:1

Reaction conditions: 1 mmol imide, 2 mmol 1-hexyne, 2 mol% Ru(met)$_2$(cod), 6 mol% n-Bu$_3$P, 4 mol% Sc(OTf)$_3$, DMF (3 mL), 60 °C, 15 h. a) isolated yield. b) ratio determined by GC. c) after 4 days reaction time. d) GC-yield with n-tetradecane as internal standard.

Cyclic imides such as succinimide (**1a**), glutarimide (**1b**) or 3,3-dimethylglutarimide (**1c**) reacted with 1-hexyne in high selectivities and the corresponding products were isolated in good yields. We obtained also very good results with the aromatic phthalimide (**1d**), although in the case of the much more electron-rich 3,4-pyridindicarboximide (**1e**), only low yields were observed. Non-cyclic imides such as diacetamide (**1f**) also reacted with 1-hexyne but only in very low yields so that the isolation of the product was not possible.

Table 8: further addition of different imides to 1-hexyne.

imide	product	yield (%)[a]	ratio (3:4)[b]
1g	3ga	90	7:1
1h	3ha	8[d]	4:1
1i	3ia	16[d]	3:1
1j	3ja	4[d,e]	3:1

Reaction conditions: 1 mmol imide, 2 mmol 1-hexyne, 2 mol% [Ru(cod)(met)$_2$], 6 mol% n-Bu$_3$P, 4 mol% Sc(OTf)$_3$, DMF (3 mL), 60 °C, 15 h. a) isolated yield. b) ratio determined by GC. c) after 4 days reaction time. d) GC-yield with n-tetradecane as internal standard. e) reaction carried out at 25 °C.

The same good results as with succinimide were obtained with 1,5,5-trimethylhydantoine (**1g**), and surprisingly the less sterically demanding 1-methylhydantoine (**1h**) only gives very poor yields, thus probably its sensibility leading to its degradation. Two other examples of cyclic imides bearing another heteroatom, reacted with 1-hexine, 2,4-thiazolidindione (**1i**) and 1-methyluracil (**1j**), although with very poor yields.

Addition of imides to alkynes

Table 9: addition of succinimide to different alkynes.

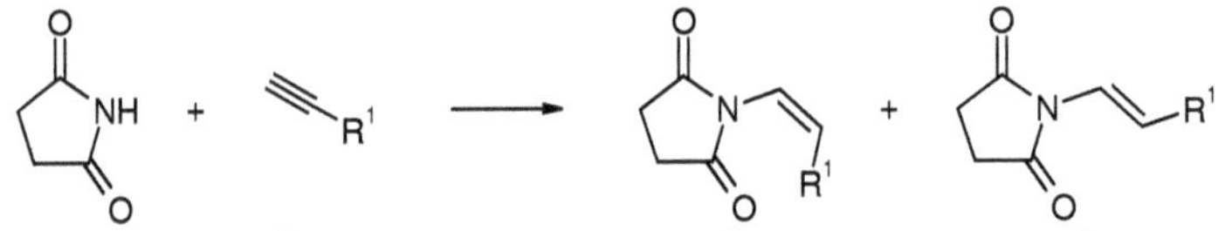

alkyne	product	yield (%)[a]	ratio (3:4)[b]
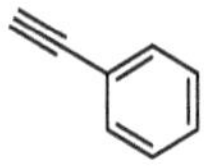 2b	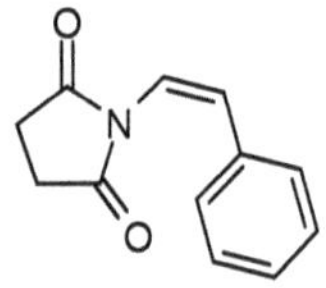 3ab	57	20:1
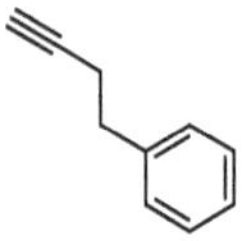 2c	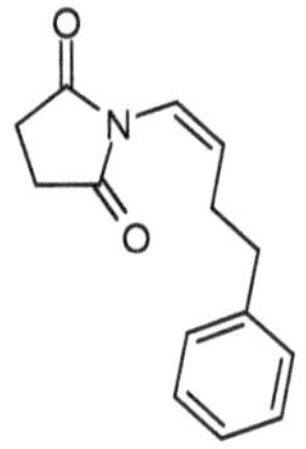 3ac	99	10:1
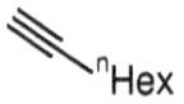 2d	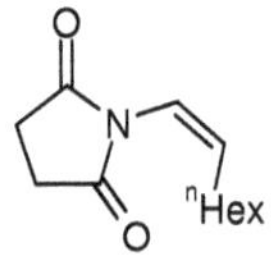 3ad	74 (77)[c]	9:1
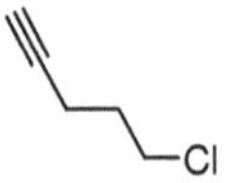 2e	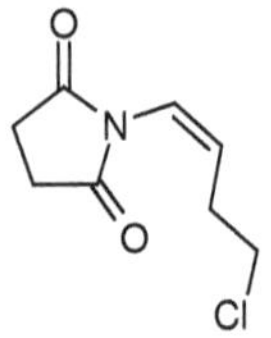 3ae	57	4:1
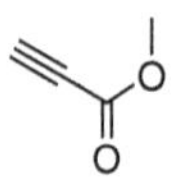	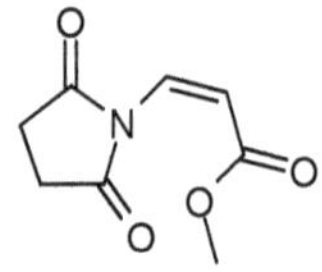	16[d]	1:1

- 58 -

Addition of imides to alkynes

alkyne	product	yield (%)[a]	ratio (3:4)[b]
2f	3af		
2g	3ag	6[d]	6:1
2h	3ah	8[d]	2:1
2i	3ai	6[d]	16:1

Reaction conditions: 1 mmol succinimide, 2 mmol alkyne, 2 mol% [Ru(cod)(met)$_2$], 6 mol% n-Bu$_3$P, 4 mol% Sc(OTf)$_3$, DMF (3 mL), 60 °C, 15 h. a) isolated yield. b) ratio determined by GC. c) after 4 days reaction time. d) GC yield with n-tetradecane as internal standard.

As shown in the table above, imide **1a** was added smoothly to different alkynes. Aliphatic (**2b**) and aromatic (**2d**) alkynes were converted into the corresponding enimides in very good yields and selectivities. Halogenides (**2e**) are also tolerated by the catalytic system. Sterically demanding alkynes such as 3,3-dimethyl-1-butine (**2g**) and olefinic alkynes (**2h**) only react in very low yields. In this case, we observed predominantly formation of a side product due to the dimerization of the alkyne. Alkynes bearing an oxygen containing functional group, such as ether (**2i**) and ester (**2f**) reacted with difficulties, probably because of a coordination of the oxygen atom to the metal instead of the triple bond. However, non protected alcohols (e.g. as 1-propynol) do not react at all.

We next extended the scope of the reaction by testing different imides with phenylacetylene (**2b**) and 4-phenyl-1-butine (**2c**).

Table 10: addition of different imides to phenylacetylene and 4-phenyl-1-butine.

imide	alkyne	product	yield (%)[a]	ratio (3:4)[b]
1d	2b	3db	53	>20:1
1g	2b	3gb	33	18:1
1c	2b	3cb	50	>20:1
1b	2b	3bb	60	>20:1

Addition of imides to alkynes

imide	alkyne	product	yield (%)[a]	ratio (3:4)[b]
1g	2c	3gc	94	>20:1
1c	2c	3cc	98	>20:1
1b	2c	3bc	88	>20:1

Reaction conditions: 1 mmol imide, 2 mmol alkyne, 2 mol% [Ru(cod)(met)$_2$], 6 mol% n-Bu$_3$P, 4 mol% Sc(OTf)$_3$, DMF (3 mL), 60 °C, 15 h. a) isolated yield. b) ratio determined by GC.

As depicted in the table above, the new catalyst system allows the addition of cyclic and aromatic imides as well as heterocyclic imides to aliphatic or aromatic alkynes in good yields.

2.1.4 Reversal of the selectivity

To complete the scope of the reaction, we developed the *E*-protocol of the reaction. Starting with the conditions of the entry 28 of the table 5 we used Sc(OTf)$_3$ instead of Yb(OTf)$_3$ which permitted us to enhance the activity of the catalyst (entry 32). Finally, after rising the catalyst loading to 5 mol %, the *E*-enimide **4** was obtained in high yield and excellent selectivity (entry 33).

Table 11: optmization of the *E*-protocol.

entry	additives	ligand	yield (%)[a]	ratio (3aa:4aa)[b]
28	Yb(OTf)$_3$	i-Pr$_3$P	35	1:15
32	Sc(OTf)$_3$	i-Pr$_3$P	48	1:15
33[c]	Sc(OTf)$_3$	i-Pr$_3$P	78	1:15

Reaction conditions: 1 mmol succinimide, 2 mmol 1-hexyne, 2 mol% [Ru(cod)(met)$_2$], 6 mol% ligand, 4 mol% Yb(OTf)$_3$ or Sc(OTf)$_3$, DMF, 100 °C, 15 h. a) corrected GC yields with *n*-tetradecane as internal standard. b) ratio determined by GC. c) reaction performed with 5 mol% [Ru(cod)(met)$_2$] and 15 mol% i-PrP$_3$.

The scope of the *E*-selective protocol was investigated using a more limited number of substrates, but always gave similarly high yields and selectivities as for the Z-protocol (Table 12).

Table 12: scope of the E-protocol.

[Reaction scheme: R¹C(O)NHC(O)R² (1) + alkyne HC≡CR³ (2) → R¹C(O)N(C(O)R²)CH=CHR³ (3) + R¹C(O)N(C(O)R²)CH=CHR³ (4)]

imide	alkyne	product	yield (%)[a]	ratio (3:4)[b]
1a (succinimide)	2a (HC≡C-ⁿBu)	4aa	75	1:15
1b (glutarimide)	2a	4ba	40	1:7
1c (dimethyl-glutarimide)	2a	4ca	28	1:7
1a	2b (phenylacetylene)	4ab	49	1:15
1a	2c (4-phenyl-1-butyne)	4ac	78	1:10

Reaction conditions: 1 mmol imide, 2 mmol alkyne, 5 mol% [Ru(cod)(met)$_2$], 15 mol% i-Pr$_3$P, 4 mol% Sc(OTf)$_3$, DMF, 100 °C, 15 h. a) isolated yield. b) ratio determined by GC.

In summary, two complementary catalyst systems have been developed that efficiently mediate the addition of imides to terminal alkynes, giving rise to either the *E*- or the *Z-anti*-Markovnikov products. The mechanistic aspect of the reaction will be discussed in a next chapter.

2.2 Addition of primary amides to alkynes

2.2.1 Strategy for the development of a new catalyst system for the addition of primary amides to alkynes

Primary amides are another particularly attractive class of N–H nucleophiles that we have not been able yet to add to alkynes using our initial catalyst system.

The addition of primary amides bears additional challenges when compared to the analogous addition of secondary amides and imides. Besides the known problems of regio- and stereoselectivity (Scheme 61: **9, 10, 15**), the product obtained from the initial addition of a primary amide to an alkyne can react with another equivalent of the alkyne to give the double addition product (Scheme 61: **11, 12, 13, 14**). Depending on the regio- and stereoselectivity of the catalyst, a mixture of up to seven different products could result from a single transformation. Therefore, it is clearly a great challenge to find a catalytic system that selectively facilitates the formation of a single mono-addition product with complete control of regio- and stereoselectivity.

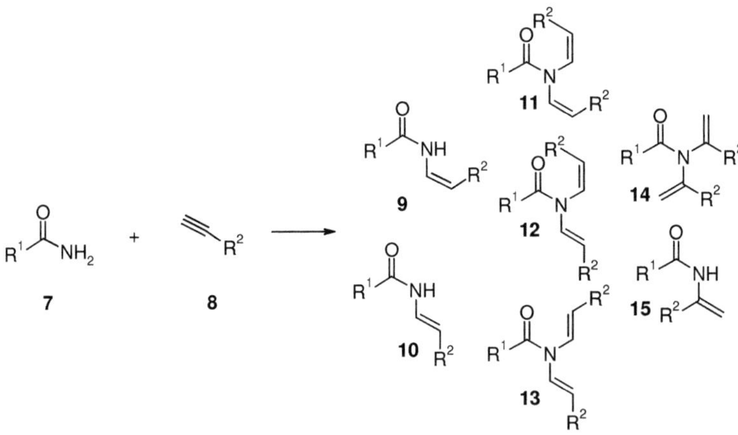

Scheme 61: possible products during the addition reaction of primary amides to alkynes.

Moreover, a perfect control of the rate and the selectivity of the first addition could allow us to perform selectively a second addition, using a second equivalent of alkynes, similar or different from the first one, giving us access to bis-enamides (Scheme 62).

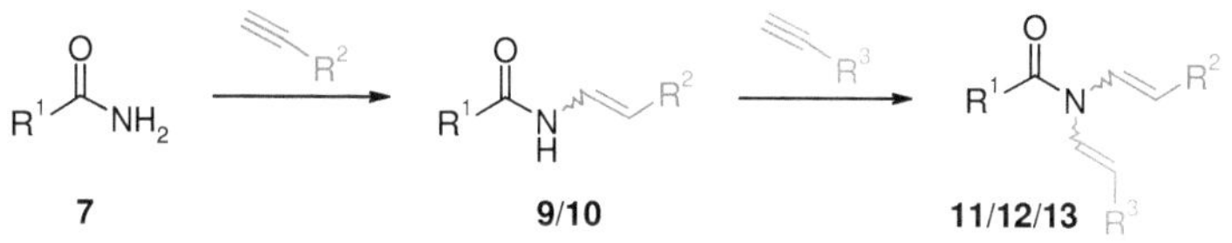

$$R^1\text{-CO-NH}_2 \quad \xrightarrow{R^2} \quad R^1\text{-CO-NH-CH=CH-}R^2 \quad \xrightarrow{R^3} \quad R^1\text{-CO-N(CH=CH-}R^3\text{)-CH=CH-}R^2$$

7   **9/10**   **11/12/13**

Scheme 62: selective single and double addition of amides to alkynes.

The inferior reactivity of primary compared to secondary amides is easily explained by their lower nucleophilicity and underlines the magnitude of the challenge posed by this substrate class: a highly developed catalyst system is required that reaches a new level of activity and stereoselectivity for the *anti*-Markovnikov addition of primary amides to alkynes but is unable to mediate the conversion of the more nucleophilic and sterically only slightly more demanding monovinylated products **9** or **10** (Scheme 62).

Initially, I was the only student on this project. For a short time, the postdoctoral researcher Iyad Karame joined in but moved to another project after his attempts to control the stereochemistry of this transformation remained unsuccessful. After my discovery that the reaction turnover can be improved by using Lewis acids, Kifah S. M. Salih joined the project and successfully developed an HPLC method which for the first time allowed the precise determination of stereoselectivities thus laying the foundation for the development of a stereoselective method. All further screening experiments, e. g. those reported in Table 13, 14, 15, 16 and 17, were carried out together, and the scope of the reaction (Table 18 and Table 19) was also investigated in close collaboration. These results thus appear in both our PhD theses.

2.2.2 Development of a new catalyst system

In order to identify a catalyst system with such a unique profile, we selected the reaction of benzamide with 1-hexyne as the model system to examine the catalytic activity of various ruthenium sources in combination with different ligands, solvents and additives (Scheme 63).

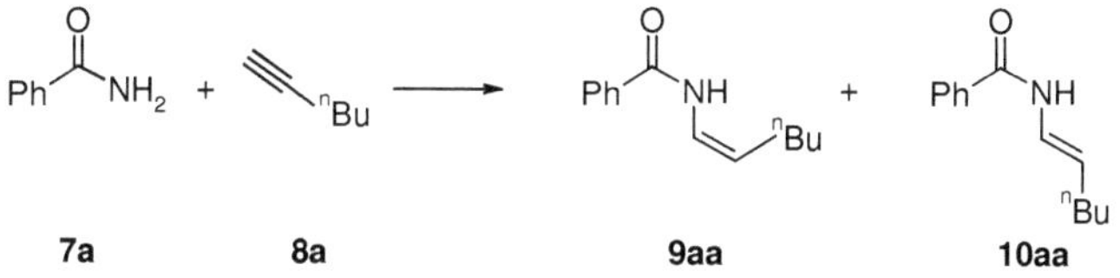

$$\text{Ph-CO-NH}_2 + \text{HC}\equiv\text{C-}^n\text{Bu} \longrightarrow \text{Ph-CO-NH-CH=CH-}^n\text{Bu} + \text{Ph-CO-NH-C(=CH}_2\text{)-}^n\text{Bu}$$

7a   **8a**   **9aa**   **10aa**

Scheme 63: addition of benzamide to 1-hexyne.

Test of different ruthenium sources

As expected, a combination of [Ru(cod)(met)$_2$] with n-Bu$_3$P and 4-(dimethylamino)pyridine (DMAP), the most effective system for the analogous reaction of secondary amides, gave only marginal conversion and displayed no selectivity for the monoaddition products (Table 13, entry 1). We first set up a range of experiments aiming to find a more active ruthenium source (Table 13).

Table 13: test of different ruthenium sources.

entry	[cat.]	yield (%)^a	ratio (9aa:10aa)^b
1	[Ru(cod)(met)$_2$]	<5	-
2	[Ru$_3$(CO)$_{12}$]	0	-
3	[Ru(acac)$_3$]	0	-
4	[RuCl$_2$(CO)$_3$]$_2$	0	-
5	[RuCl$_2$(C$_6$H$_6$)]$_2$	0	-
6	[(p-cymene)RuCl$_2$]$_2$	2	-
7	[AuCl$_3$]	0	-

Reaction conditions: 1 mmol benzamide, 2 mmol 1-hexyne, 5 mol% Ru-source or AuCl$_3$, 15 mol% n-Bu$_3$P, 4 mol% DMAP, toluene, 100 °C, 15 h. a) corrected GC yields with n-tetradecane as internal standard. b) ratio determined by HPLC.

Other ruthenium precursors were found to be even less efficient (entries 2-6), and even gold catalysts which are very useful catalyst in alkyne chemistry did not show any reactivity (entry 7).

The influence of the solvent

As the choice of the solvent appears to be fundamental in the addition of imides to alkynes we then decided to perform the reaction in different solvents (Table 14).

Table 14: test of different solvents.

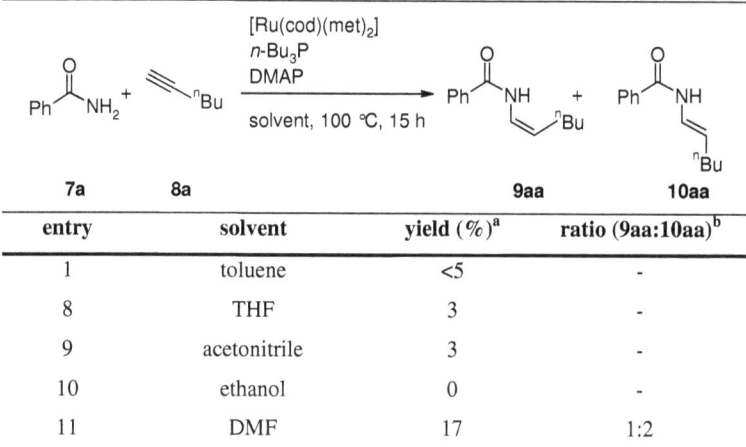

entry	solvent	yield (%)[a]	ratio (9aa:10aa)[b]
1	toluene	<5	-
8	THF	3	-
9	acetonitrile	3	-
10	ethanol	0	-
11	DMF	17	1:2

Reaction conditions: 1 mmol benzamide, 2 mmol 1-hexyne, 5 mol% Ru(met)$_2$(cod), 15 mol% n-Bu$_3$P, 4 mol% DMAP, solvent, 100 °C, 15 h. a) corrected GC yields with n-tetradecane as internal standard. b) ratio determined by HPLC.

Protic solvents such as ethanol (entry 10) appear to be totally inefficient. Acetonitrile, probably because of coordination to the ruthenium center slowed down the reaction (entry 9). DMF was uniquely effective as for the addition of imides to alkynes, resulting in the first increase in yield (entry 11).

The influence of the additives

As experienced during our work on the addition of secondary amides and imides to alkynes, the reactivity of amides in regards to alkynes seems to be very dependent on the additive used. We then next tested different additives.

Table 15: test of different additives.

entry	additive	yield (%)[a]	ratio (9aa:10aa)[b]
11	DMAP	17	1:2
12	Na_2CO_3	16	1:3
13	KI	17	1:1
14	LiOTf	<5	-
15	$Mg(OTf)_2$	45	2:1
16	$Cu(OTf)_2$	0	-
17	$Zn(OTf)_2$	0	-
18	$Bi(OTf)_3$	<5	-
19	$Sc(OTf)_3$	35	1:1
20	$Yb(OTf)_3$	59	3:2

Reaction conditions: 1 mmol benzamide, 2 mmol 1-hexyne, 5 mol% [Ru(cod)(met)$_2$], 15 mol% n-Bu$_3$P, 4 mol% additive, DMF, 100 °C, 15 h. a) corrected GC yields with n-tetradecane as internal standard. b) ratio determined by HPLC.

The use of inorganic bases such as sodium carbonate (entry 12) did not lead to any improvement. Based on the reasoning that the coordination of a Lewis acid to the carbonyl oxygen would acidify the amide protons and therefore serve the same purpose as an added base, we replaced DMAP, which we had found to aid the deprotonation of the amide substrate and facilitate its coordination to the Ru-center, by a Lewis acid (entries 14-20). This indeed led to a dramatic increase in catalyst activity, and 59% yield was achieved with Yb(OTf)$_3$ (entry 20). The other triflates we tested (entries 14-20) never reached this activity.

The influence of the ligands

To achieve the completion of the reaction and push the selectivity in favor of one of the stereoisomers, we decided to test some ligands.

Table 16: test of different ligands.

entry	ligand	yield (%)^a	ratio (9aa:10aa)^b
15	n-Bu$_3$P	59	3:2
21	N(n-Bu)$_3$	0	-
22	phenanthroline	0	-
23	PPh$_3$	10	1:1
24	P(i-Pr)Ph$_2$	14	2:1
25	PCy$_3$	0	-
26	i-Pr$_3$P	0	-
27	dcypb	55	4:1

Reaction conditions: 1 mmol benzamide, 2 mmol 1-hexyne, 5 mol% [Ru(cod)(met)$_2$], 15 mol% monodendate ligand or 6 mol% bidentate ligand, 4 mol% Yb(OTf)$_3$, DMF, 100 °C, 15 h. a) corrected GC yields with n-tetradecane as internal standard. b) ratio determined by HPLC.

The use of nitrogen ligands such as tri-n-butylamine or phenanthroline, resulted in no yields (entries 21, 22). Electron poor phosphines, such as tri-phenylphosphine, appear to slow down the reaction (entries 23, 24). Surprisingly, tri-cyclohexyl phosphine and tri-isopropyl phosphine did not show any activity (entries 25, 26), probably because of their too low stability. Finally, we found that sterically demanding, electron-rich chelating phosphines resulted in a major enhancement of the Z/E-ratio. Using 1,4-bis-(dicyclohexylphosphino)butane (dcypb), the selectivity for the Z-product **9aa** was increased to a 4:1 ratio (entry 27).

Optimization of the reaction conditions

Due to the high activity of the catalyst system, the reaction temperature could be reduced to 60 °C (Table 17, entry 29). This resulted in significantly better yields of the monoaddition products because the main side reaction, the hydrolysis of the enamide products by traces of water, no longer played an important role. At this moderate temperature, the presence of water was even found to enhance the catalyst performance (entry 31), so that finally nearly complete conversion and an impressive 18:1 selectivity for the thermodynamically disfavored and thus more interesting Z-enamide **9aa** was reached within six-hour reaction time (entry 32).

Table 17: optimization of the reaction conditions.

entry	solvent	temperature (°C)	yield (%)^a	ratio (9aa:10aa)^b
27	DMF	100	55	4:1
28	"	80	82	5:1
29	"	60	95	12:1
30	"	25	<5	-
31	DMF/water^c	60	98	18:1
32^d	"	"	97	18:1

Reaction conditions: 1 mmol benzamide, 2 mmol 1-hexyne, 5 mol% Ru(met)$_2$(cod), 6 mol% dcypb, 4 mol% Yb(OTf)$_3$, DMF, 15 h. a) corrected GC yields with *n*-tetradecane as internal standard. b) ratio determined by HPLC. c) water as co-solvent (108 µL). d) 6 h reaction time.

2.2.3 Scope of the reaction

To investigate the scope of this new hydroamidation protocol, we applied it to the addition of a broad variety of nitrogen nucleophiles to several terminal alkynes.

Table 18: addition of benzamide to different alkynes.

alkyne	product	yield (%)[a]	ratio (9:10)[b]
8a	9aa	94	18:1
8b	9ab	97	18:1
8c	9ac	91	>40:1
8d	9ad	92	17:1
8e	9ae	86	23:1

Addition of primary amides to alkynes

Ph-C(=O)-NH₂ + ≡-R¹ → Ph-C(=O)-NH-CH=CH-R¹ (branched) + Ph-C(=O)-NH-CH=CH-R¹ (linear)

7a   **8**   **9**   **10**

alkyne	product	yield (%)[a]	ratio (9:10)[b]
8f (4-chlorobut-1-yne, CH≡C-CH₂CH₂-Cl)	9af	91	31:1
8g (4-fluorophenylacetylene)	9ag	99	20:1
8h (4-methoxyphenylacetylene)	9ah	83	19:1

Reaction conditions: 1 mmol benzamide, 2 mmol alkyne, 5 mol% [Ru(cod)(met)₂], 6 mol% dcypb, 4 mol% Yb(OTf)₃, DMF (3 mL), 60 °C, 15 h. a) isolated yield. b) ratio determined by HPLC.

As shown in the table above, benzamide reacted with a wide range of alkynes, including the standard 1-hexyne (**8a**), haloalkyl (**8f**), alicyclic (**8e**) or 4-phenyl-1-butyne (**8d**), but also with sterically hindered alkynes such as phenylacetylene (**8b**), phenylacetylene analogous (**8g**, **8h**) or *tert*-butylacetylene (**8c**) yielding the corresponding products in high yields and excellent selectivities.

Furthermore, a range of primary amides, among them sensitive and highly functionalized derivatives, were smoothly added to phenylacetylene.

Table 19: addition of different primary amides to phenylacetylene.

amide	product	yield (%)[a]	ratio (9:10)[b]
7b	9bb	46	30:1
7c	9cb	88	10:1
7d	9db	90	19:1
7e	9eb	96	22:1
7f	9fb	84	14:1

amide	product	yield (%)[a]	ratio (9:10)[b]
7g (3-O2N-C6H4-C(O)NH2)	9gb	80	20:1
7h (4-F-C6H4-C(O)NH2)	9hb	93	19:1
7i (4-MeO-C6H4-C(O)NH2)	9ib	96	18.1
7j (EtO-C(O)-C(O)NH2)	9jb	78[c]	35:1
7k (N≡C-C(O)NH2)	9kb	31	>40:1

Addition of primary amides to alkynes

amide	product	yield (%)[a]	ratio (9:10)[b]
7l (EtO-CO-CH2-CO-NH2)	9lb (EtO-CO-CH2-CO-NH-CH=CH-Ph)	71[c]	19:1

Reaction conditions: 1 mmol amide, 2 mmol phenylacetylene, 5 mol% [Ru(cod)(met)$_2$], 6 mol% dcypb, 4 mol% Yb(OTf)$_3$, DMF (3 mL), 60 °C, 15 h. a) isolated yield. b) ratio determined by HPLC. c) the reaction was performed without water to avoid hydrolysis of the sensitive products.

Common functional groups such as dialkylamine (**7b**), *tert*-butyl (**7c**), alkene (**7d**), acetyl (**7e**), alkane (**7f**), nitro (**1g**), and halide groups (**7h**) were tolerated, and even fragile oxalic (**7j**), malonic (**7l**) or α-cyanoacetic (**7k**) amides were smoothly converted. Arguably the most spectacular result was obtained in the synthesis of enamide **9eb**, where a primary amide functionality was selectively vinylated in the presence of a secondary amide group.

2.2.4 Reversal of the selectivity

We next tried to invert the selectivity of the reaction in favor of the *E*-enamide **9** by modifying the catalyst system, and thus to further improve the synthetic utility of the hydroamidation protocol. Disappointingly, the best result, obtained with *n*-Bu$_3$P as the ligand and 3 Å molecular sieves as additives, was a 2.5:1 ratio of *E*- to *Z*-isomer. A superior strategy proved to be isomerizing the double bond isomers *in situ* after the hydroamidation reaction was complete, by adding triethylamine and heating the reaction mixture to 110 °C while using molecular sieves to reduce the hydrolytic cleavage to a minimum (Scheme 64).

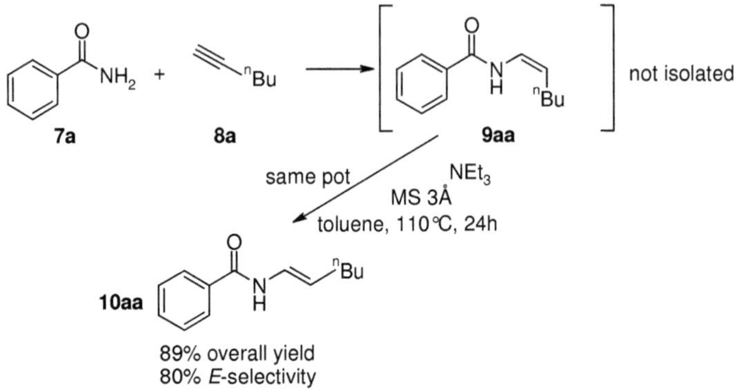

Addition of primary amides to alkynes

Scheme 64: one pot Z-hydroamidation of 1-hexyne followed by isomerization.

This way, an E/Z-ratio of 4:1 was achieved for the model system, and as can be seen from the table below, E-selectivities of up to 20:1 were reached for other substrates.

Table 20: scope for the E-protocol of the addition of primary amides to alkynes.

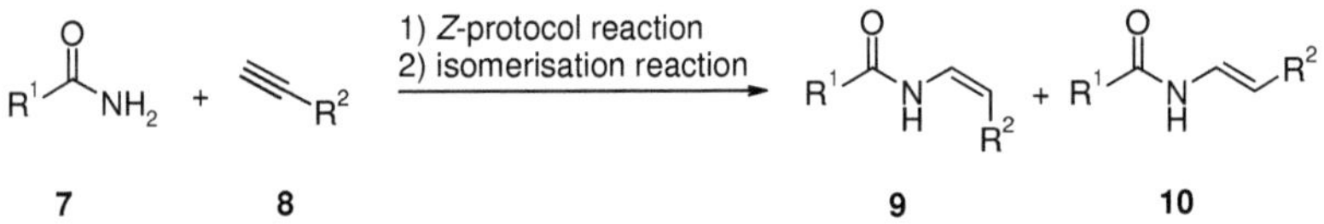

amide	alkyne	product	yield (%)[a]	ratio (9:10)[b]
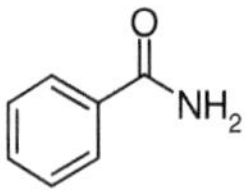 7a	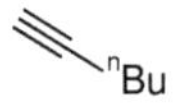 8a	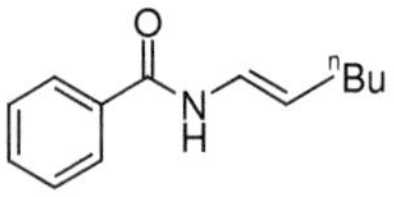 10aa	89	1:4
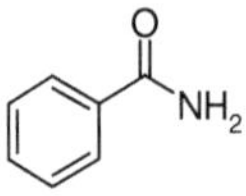 7a	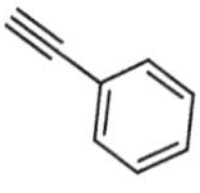 8b	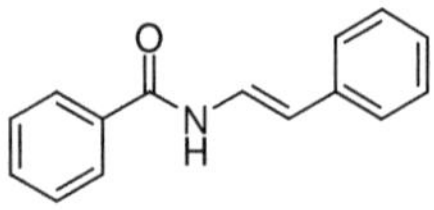 10ab	92	1:18
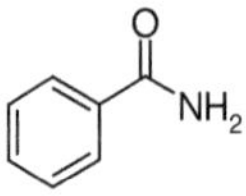 7a		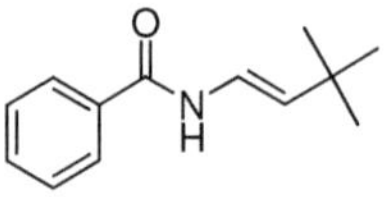	79	1:20

Addition of primary amides to alkynes

amide	alkyne	product	yield (%)[a]	ratio (9:10)[b]
7a	8c	10ac	85	1:4
7a	8d	10ad	85	1:4
7a	8e	10ae	83	1:17
7d	8b	10db	77	1:20
7e	8b	10eb	68	1:14
7f	8b	10fb		

Reaction conditions: 1 mmol amide, 2 mmol alkyne, 5 mol% [Ru(cod)(met)$_2$], 6 mol% dcypb, 4 mol% Yb(OTf)$_3$, DMF (3 mL), 60 °C, 15 h. After complete reaction, 500 mg of 3 Å molecular sieves and 200 μL triethylamine were added, and the mixture was heated to 110 °C for 24 h. a) isolated yield. b) ratio determined by HPLC.

Enamides thus accessible in only one step from easily available materials include the natural products lansiumamide A (**9mb**)[6] and alatamide (**10ah**) (Scheme 65),[77] whose previously published syntheses involved three to four steps and gave inferior yields, and compound **9nh**, which is a key intermediate in Castedo´s total synthesis of aristolactam (Scheme 65).[78]

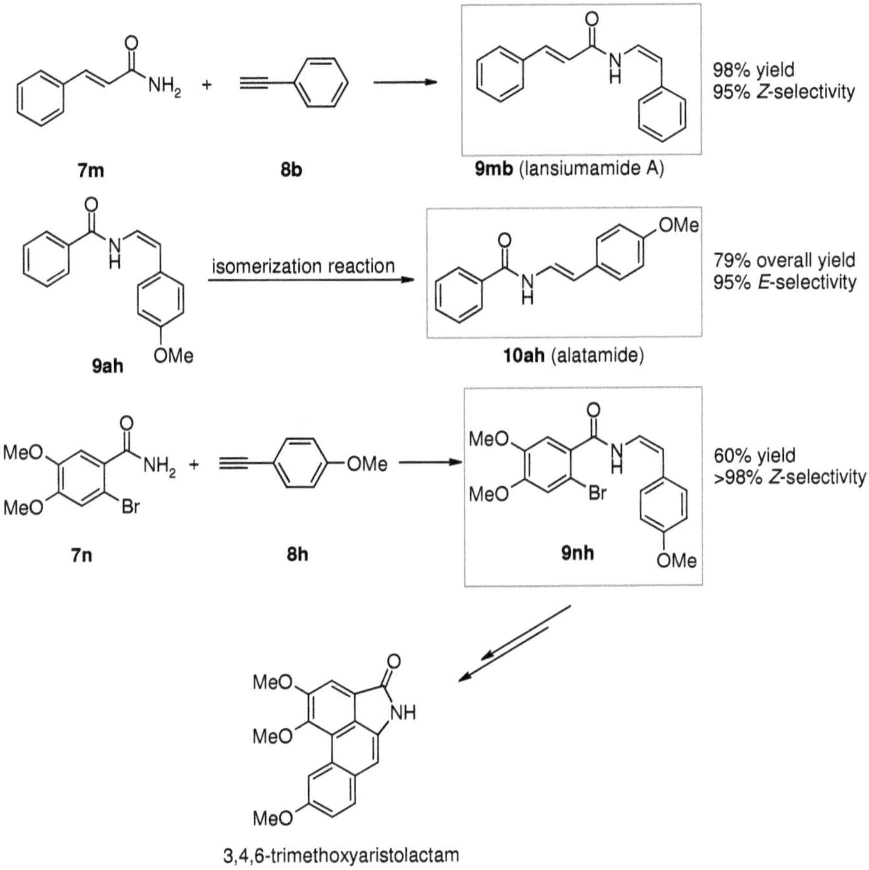

Scheme 65: synthesis of some natural products.

2.2.5 Mechanistic hypothesis

Our mechanistic proposal for such hydroamidations is that the alkyne first coordinates to a Ru^{II}-amide complex generated from the catalyst precursor, then an amide anion adds to the C-C-triple bond giving rise to a η^1-Ru-vinyl complexes (**c, d**). Depending on whether the attack occurs from inside (**c**) or outside (**d**) the coordination sphere of the ruthenium, the metal will end up Z- or E- to the amide group. The product is liberated by protonolysis, regenerating the initial Ru^{II}-species (**a**) and completing the catalytic cycle (Scheme 66).

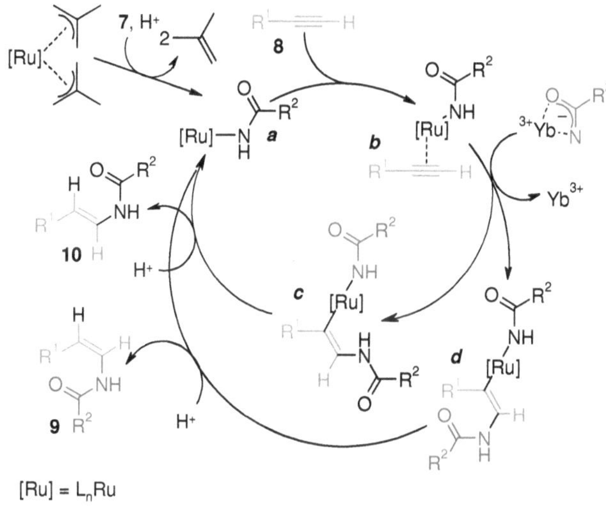

Scheme 66: mechanistic hypothesis for the addition of primary amides to alkynes.

Mechanistic studies according to this transformation and based on observations made during our work on the addition of secondary amides and imides to alkynes will be describe in a next chapter.

2.2.6 Second addition

As mentioned earlier, after addition of an amide to a first equivalent of alkyne, the product formed can react with a second equivalent of alkyne. However, the activity and selectivity obtained with our system permits us to stop the reaction after the first addition. We then tried, after isolation of this first addition product, to put the secondary enamide with a second equivalent of alkyne under conditions similar to those used for the addition of secondary addition to alkynes developed by Gooßen and Rauhaus (Scheme 67).

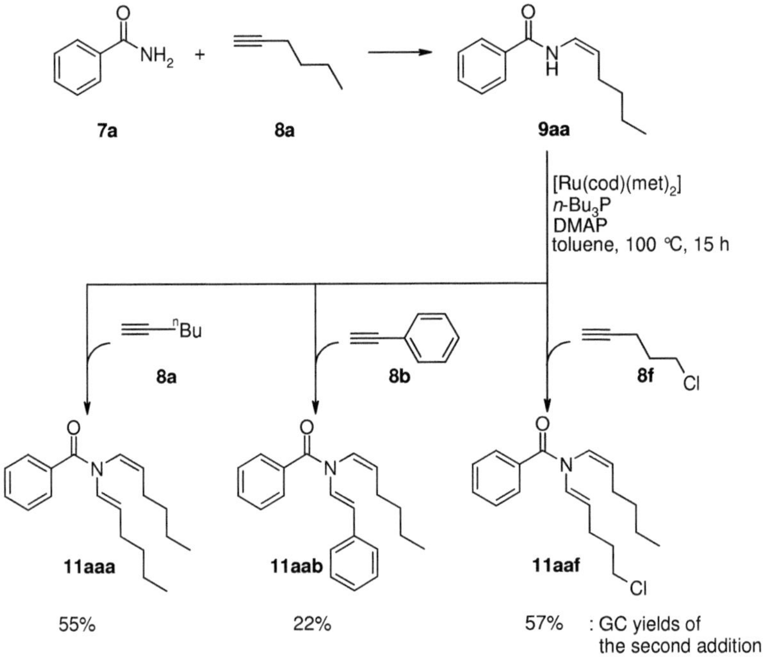

Scheme 67: first and second selective addition of amides to alkynes.

We were pleased to find that these first tests already gave fair isolated yields; indeed, after the first addition of benzamide to 1-hexyne we were able to add selectively another equivalent of 1-hexyne, but also phenylacetylene and 5-chloropentyne to the secondary enamide formed.

This new reaction not yet well optimized is still very promising, and new compounds and a completely new family of compounds could be synthesized using this methodology. Among them, cyclization product, by using diynes could lead to very interesting products (Scheme 68)

Scheme 68: possible synthesis using the double addition of amides to alkynes.

2.2.7 Conclusion

Overall, a new catalyst system has been developed that allows the chemo-, regio-, and stereoselective synthesis of secondary Z-enamides from easily available primary amides and terminal alkynes. In combination with an optional *in situ* isomerization to the E-enamides, this represents an expedient synthetic entry to this structural motif, which is featured in many natural products and is hard to access by other means. The mechanism of the reaction and particularly the role of the triflate salts is not yet well understood, but the mechanistic studies currently in progress may clarify this aspect.

2.3 Second generation catalyst for the addition of secondary amides to alkynes

2.3.1 Mechanistic studies

As shown in the two last chapters, the Ru-catalyzed addition of amides to alkynes is a very powerful reaction. It allows the synthesis of numerous enamides in high yields and selectivities. However, practical applications of this ruthenium-catalyzed additions have so far been hampered by the prohibitive price of ruthenium(II) complexes stabilized with labile ligands, e.g. [Ru(cod)(met)$_2$] (68 €/mmol)[79], that is required for generating the catalytically active species and by the use of highly air sensitive and pyrophoric n-Bu$_3$P. The replacement of this catalytic system by an easier to handle phosphine and a readily available ruthenium source such as RuCl$_3\cdot$3 H$_2$O (5 €/mmol)[79] could vastly improve the synthetic utility of this transformation. Even if such a replacement could be difficult, considering the necessity of selectively reducing ruthenium(III) to ruthenium(II) during the reaction, it should still be feasible since RuCl$_3\cdot$n H$_2$O is the starting material of a wide scope of ruthenium complexes, including [Ru(cod)(met)$_2$] (Scheme 69).[80]

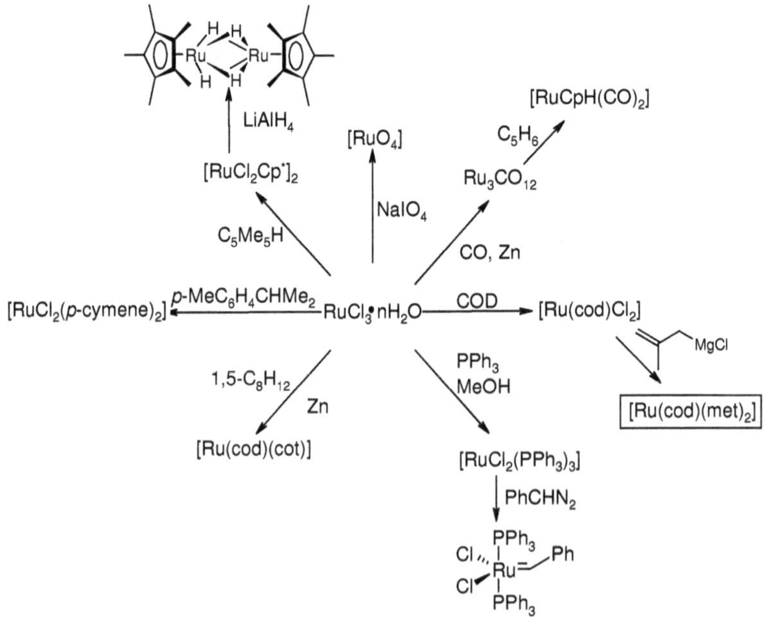

Scheme 69: different syntheses of complexes starting from RuCl$_3\cdot$n H$_2$O.

Second generation catalyst for the addition of secondary amides to alkynes

This project was carried out in collaboration with Matthias Arndt and Felix Rudolphi, who both focused on the synthesis of new ruthenium precusors and on the in situ generation of active catalysts. My share was the development of the hydroamidation conditions suitable for the new systems. Our combined efforts led to the discovery of the new reaction protocol detailed below. We jointly isolated the compounds Table 24 in order to demonstrate the scope of this protocol. The results were published in *Advanced Synthesis and Catalysis*.[81]

Because of the wide scope of the reaction and the relative simplicity of the catalytic system, we chose the addition of secondary amides to alkynes as a starting point for our investigations. Our first objective was to pinpoint the active species engaged in the catalyzed reaction between pyrrolidin-2-one and 1-hexyne (Scheme 70).

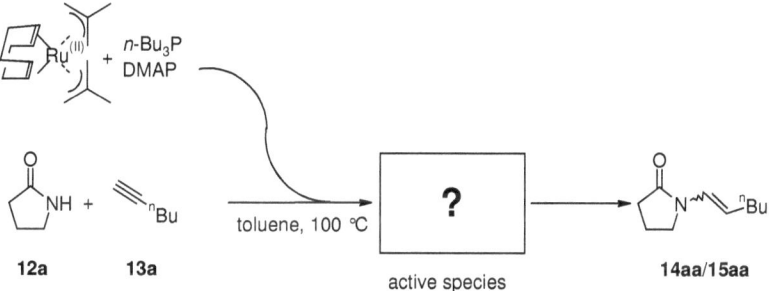

Scheme 70: formation of an active species during the reaction between pyrrolidin-2-one and 1-hexyne.

ESI-MS (Electrospray Ionisation-Mass Spectroscopy) of a sample taken directly from the reaction solution during the catalyst preformation displayed two predominant signals (Scheme 71). One signal at m/z = 712.3 which corresponds to $[(n\text{-}Bu_3P)_2Ru(C_4H_6NO)(C_7H_{10}N_2)]^+$ and another signal at m/z = 792.4 which could be assigned to $[(n\text{-}Bu_3P)_3Ru(C_4H_6NO)]^+$. Additionally, standard GC-MS of this sample also shows signals for free COD and isobutene.

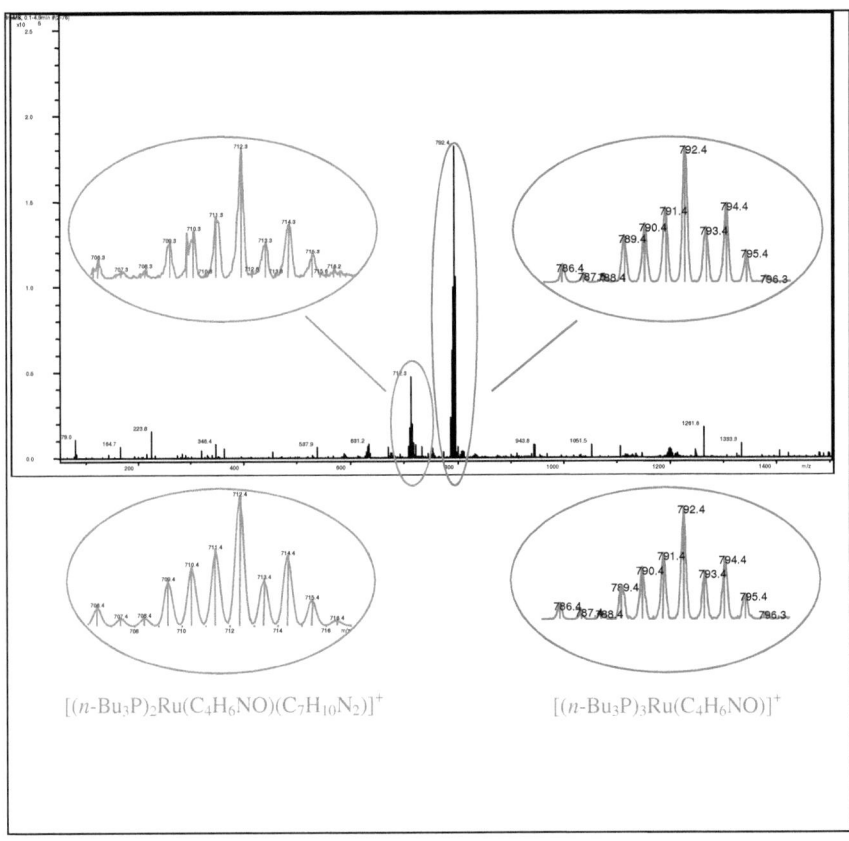

Scheme 71: ESI-MS of the catalyst systems based on [Ru(cod)(met)$_2$].

Second generation catalyst for the addition of secondary amides to alkynes

The *in situ* formation of the catalyst from [Ru(cod)(met)$_2$] was investigated by a series of ^{1}H- and ^{31}P-NMR experiments (Scheme 72 and Scheme 73).

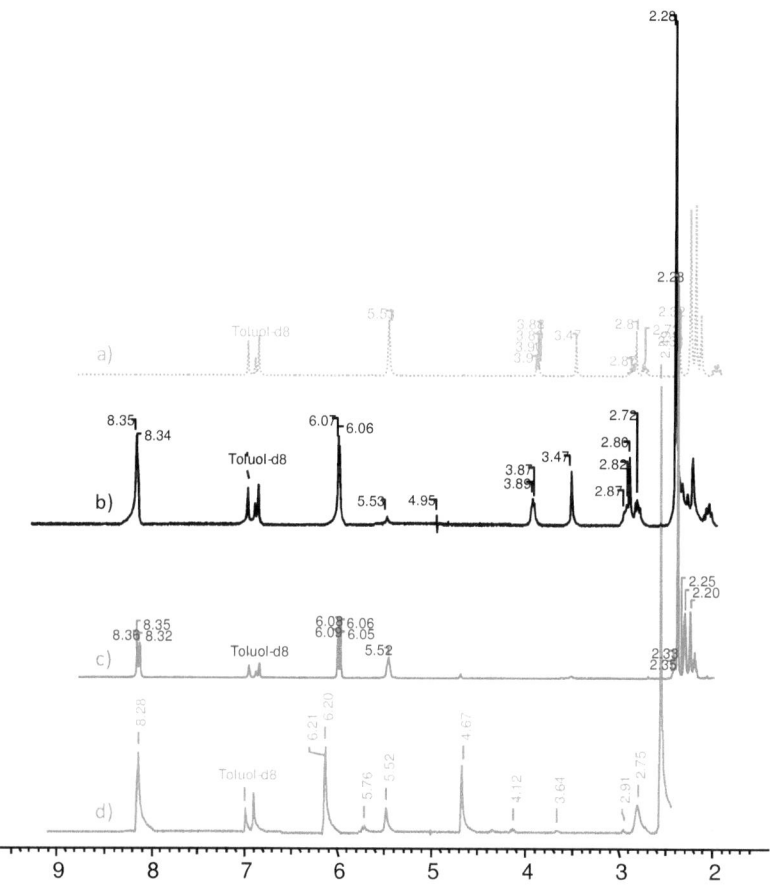

Scheme 72: ^{1}H-NMR of the catalyst preformation.

A 5 mm NMR tube was charged with [Ru(cod)(met)$_2$] (12.8 mg, 0.04 mmol). The tube was sealed with a rubber septum and purged with alternating vacuum and nitrogen cycles. Tri-*n*-butylphosphine (30 µL, 0.012 mmol) and toluene-d_8 (1 mL) were added via syringe. For clearness reasons, the section $\delta < 2.20$ ppm of the ^{1}H-NMR spectra (in which the alkyl signals corresponding to the excess of phosphine appears) was blended. The ^{1}H-NMR spectrum taken at this point (25 °C), prior to catalyst preformation (**a**), shows clear singlets at 3.47 (2H), 2.80

- 85 -

(2H), 1.69 (6H), 1.53 (2H), and 0.15 (2H) ppm for two ruthenium-bound methallyl ligands, and multiplets at 3.87-3.92, 2.82-2.91, 2.68-2.75, 1.86-1.97 ppm (2H each), 1.55-1.63 and 1.09-1.16 ppm (2H each, overlapping with signals of the phosphine ligands) for the ruthenium-bound COD ligand.

A second sample was prepared under the same conditions but with a two-fold excess of DMAP. A first ^{1}H-NMR taken at this point at 25°C (**b**) displayed the same signals as before plus the signals of DMAP at 8.34 (2H), 6.07 (2H) and 2.28 (6H). After heating the reaction mixture to 100 °C for 5 min, another ^{1}H-NMR was taken (**c**). COD ligand had completely been displaced. In the ^{1}H-NMR spectrum, the signals for coordinated COD ligands have disappeared, and signals at 5.43-5.57 (4H) and 2.14-2.24 (8H) ppm indicate free 1,5-cyclooctadiene. Due to the ligand exchange, the signals for the methallyl ligands are shifted to higher field, making an assignment difficult. Signals for isobutene which would indicate protonolysis or decomposition of the ruthenium complex cannot be observed. After the addition of pyrrolidin-2-one (6.2 µL, 0.08 mmol) via syringe, the methyl ligands were protonated off by the amide, and signals at 4.60-4.67 and 1.61 ppm in the ^{1}H-NMR spectrum (**d**) gave evidence of the formation of isobutene. The α-protons of the amide were found at 2.62-2.80 ppm, whereas the other ring protons are concealed among the overlapping alkyl signals of tri-*n*-butylphosphine in the range of 0.0-2.5 ppm.

The ^{31}P-NMR (Scheme 73) spectrum at 25 °C shows a minor signal at 20.8 ppm for metal-coordinated tri-n-butylphosphine[82] and a strong signal at 31.1 ppm for uncoordinated phosphine (**b**).[83] After heating the above solution to 100 °C for 5 min (**c**), a strong signal for coordinating phosphine appeared at 20.8 ppm, meaning that most of the phosphine was bond to the ruthenium center at this point of the reaction. Finally, after addition of the pyrrolidin-2-one, only a signal corresponding to the coordinated phosphine was displayed (**d**).

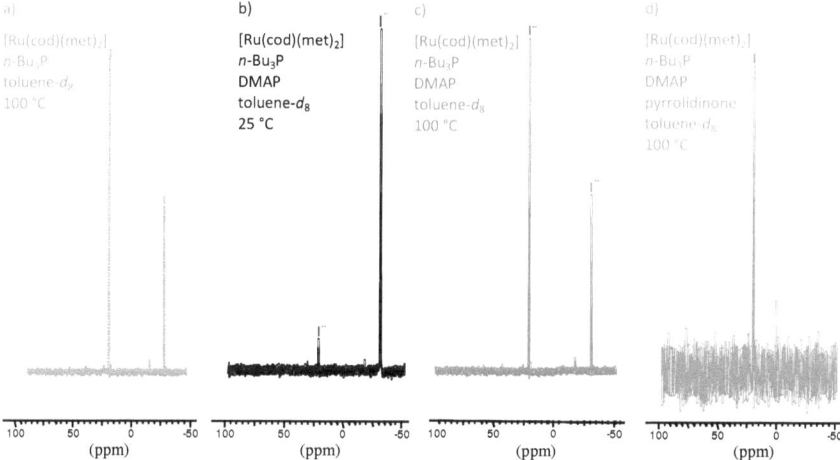

Scheme 73: ^{31}P-NMR of the catalyst preformation.

To summarize (Scheme 74), during the preformation steps, the 1,5-cyclooctadiene (COD) ligand is displaced by the phosphine and/or DMAP ligands giving rise to intermediate **b**, as indicated by the characteristic signals for free, unaltered COD and Ru-coordinated n-Bu$_3$P in GC-MS / NMR spectra. Moreover, the detection of isobutene suggests that the methallyl ligands are protonated off by the acidic N–H of the amides, thus leading to a Ru(II) amide complex **c**.

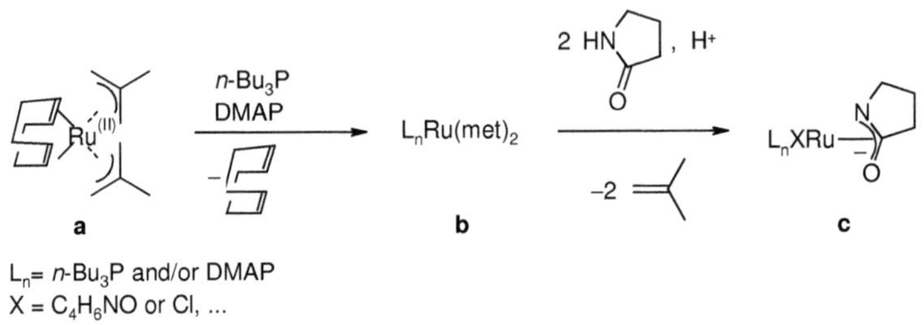

Scheme 74: suggested catalyst preformation.

L_n = n-Bu$_3$P and/or DMAP
X = C$_4$H$_6$NO or Cl, ...

This results show that none of the original ligands remained actually bound to ruthenium center, suggesting that simple ruthenium salts such as RuCl$_3$·3 H$_2$O could also be utilized. By searching for a method to reduce RuCl$_3$·3 H$_2$O to Ru(II) while at the same time removing the chloride ions, which can be expected to compete with the substrate for coordination to the ruthenium center, we came across a publication by Kölle et al.[84] They described the synthesis of di-μ-methoxo-bis[(η^5-pentamethylcyclopentadienyl) diruthenium(II) [(Cp*RuOMe)$_2$] from di-μ-chloro-bis[(η^5-pentamethylcyclopentadienyl)chlororuthenium(III)] [(Cp*RuCl$_2$)$_2$] using MeOH as a reductant and the mild base K$_2$CO$_3$ for the generation of the methoxide ligands as well as for precipitation of the chloride as KCl.

The first test using these conditions on our test reaction was very successful (Scheme 75 and Table 21, entry 1). Indeed, the desired E-N-(hex-1-enyl)pyrrolidin-2-one product had formed in high (E/Z)-selectivity and reasonable but varying yields.

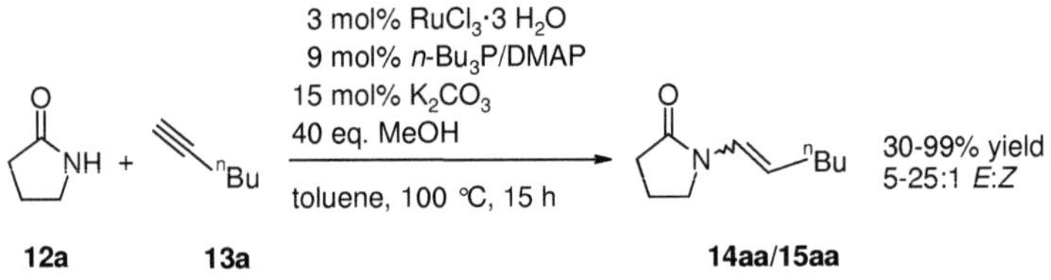

Scheme 75: first attempt using RuCl$_3$·3 H$_2$O as catalyst for our test reaction.

Nevertheless, in our search for a better protocol, we made an interesting observation. To facilitate our task, we prepared a stock solution of the catalyst. I.e. if the catalyst is soluble in the reaction solvent, we dissolved it and thus it became easier to always add the same amount

Second generation catalyst for the addition of secondary amides to alkynes

of catalyst to the reaction vessel. However, the commercial $RuCl_3 \cdot 3\ H_2O$ was not soluble in toluene, so we decided to first dissolve it in acetone, and after evaporation of the latter from the reaction vessel, adding the remaining compounds including the solvent. The catalyst became then totally soluble in toluene. We do not fully understand the reason for this behavior. To the best of our knowledge, two modifications of $RuCl_3$ have been reported: β-$RuCl_3$ which is characterized chemically by strong hygroscopic properties and good solubility in water and polar organic solvents, and α-$RuCl_3$, which is highly inert and only soluble in hot *aqua regia*.[85]

Anyway, this peculiar behavior we observed not only enhances the solubility of $RuCl_3 \cdot 3\ H_2O$ in toluene, but also makes the reaction reproducible (Table 21, entry 2).

2.3.2 Development of a new catalyst system

We were now able to begin our investigation. The yields of the reaction already being high, we focused our efforts to make this reaction easier to handle and less expensive.

Test of different ligands

One of the primary objectives of this work was to find a less air sensitive phosphine (Table 21).

Table 21: test of different ligands.

entry	ligand	yield (%)[a]	ratio (14aa:15aa)[b]
1[c]	n-Bu$_3$P	30–99	5-25:1
2	n-Bu$_3$P	>99	18:1
3	n-Oct$_3$P	>99	19:1
4	i-Pr$_3$P	11	1.3:1
5	t-Bu$_3$P	4	3:1
6	PPh$_3$	21	2:1
7	P(o-Tol)$_3$	27	1.6:1
8	P(p-C$_6$H$_4$OMe)$_3$	17	2.4:1

entry	ligand	yield (%)[a]	ratio (14aa:15aa)[b]
9	P(2-Fur)$_3$	11	2:1
10	dcypb	2	3:1
11	Cy$_2$P(CH$_2$)PCy$_2$	6	1:1
12	[HP(n-Bu)$_3$]$^+$[BF$_4$]$^-$	>99	22:1

Reaction conditions: 1.00 mmol pyrrolidin-2-one, 2.00 mmol 1-hexyne, 2 mol% RuCl$_3$·3 H$_2$O*, 6 mol% ligand (3 mol% for chelating ligands), 4 mol% DMAP, 0.30 mmol MeOH, 10 mol% K$_2$CO$_3$, 3 mL toluene, 100 °C, 15 h; a) corrected GC yields with n-tetradecane as internal standard. b) ratio determined by GC. c) no previous dissolution in acetone was performed for this reaction.
* The catalyst was injected as a stock solution of RuCl$_3$·3H$_2$O (0.02 mmol) in acetone (1 mL) via syringe. The solvent was removed under vacuum before any liquid compounds were added.

Best results were obtained with n-Bu$_3$P and n-Oct$_3$P (entries 2, 3). In contrast, other alkylphosphines such as i-Pr$_3$P or t-Bu$_3$P did not show comparable activities, probably because of their high steric hindrance (entries 4, 5). Moderate results were observed with arylphosphines such as PPh$_3$, (o-Tol)$_3$P and P(p-C$_6$H$_4$OMe)$_3$ (entries 5, 6, 7 respectively) which is interesting because of the very low price of PPh$_3$ and its high stability. Finally, chelating phosphines did not show any activity (entries 10, 11). Full conversion with high stereoselectivity was also obtained with the tetrafluoroborate salt of n-Bu$_3$P (entry 12),[86] which is air- and water-stable, and advantageous especially for large-scale applications.

Test of different additives

Systematic investigations revealed that a base, for example potassium carbonate or hydroxide, is essential, whereas the alcohol can be replaced by water (Table 22, entries 13-16). This goes along with reports in the literature that phosphines themselves can act as reducing agents for transition metals, a process assisted by water.[87]

Table 22: variation of additives and bases.

entry	additive	base	yield (%)[a]	ratio (14aa:15aa)[b]
2[c]	MeOH	K_2CO_3	>99	22:1
13	MeOH	-	7	3:2
14	-	K_2CO_3	95	14:1
15	EtOH	K_2CO_3	>99	23:1
16	H_2O	K_2CO_3	>99	24:1

Reaction conditions: 1.00 mmol pyrrolidin-2-one, 2.00 mmol 1-hexyne, 2 mol% $RuCl_3 \cdot 3H_2O$*, 6 mol% n-Bu$_3$P, 4 mol% DMAP, 0.30 mmol additive, 10 mol% base, 3 mL toluene, 100 °C, 15 h; a) corrected GC yields with n-tetradecane as internal standard. b) ratio determined by GC.
* The catalyst was injected as a stock solution of $RuCl_3 \cdot 3H_2O$ (0.02 mmol) in acetone (1 mL) via syringe. The solvent was removed under vacuum before any liquid compound was added.

Optimization of the reaction conditions

Finally a range of experiments was set up to find out what are the ideal conditions for this reaction. It was important to observe if toluene could be replaced by a different solvent and if the temperature of the reaction could be lowered (Table 23).

Table 23: optimization of the reaction conditions.

entry	solvent	temperature (°C)	yield (%)[a]	ratio (14aa:15aa)[b]
2	toluene	100	>99	>20:1
17	PhCl	"	98	19:1
18	chinoline	"	>99	14:1
19	n-octane	"	97	>20:1
20	EtOH	"	12	4:1
21	acetonitrile	"	11	1.6:1
22	DMF	"	80	15:1
23	toluene	80	>99	>20:1
24	"	70	94	>20:1
25	"	60	37	>20:1
26	"	50	9	>20:1

Reaction conditions: 1.00 mmol pyrrolidin-2-one, 2.00 mmol 1-hexyne, 2 mol% $RuCl_3 \cdot 3 H_2O$*, 6 mol% n-Bu$_3$P, 4 mol% DMAP, 0.30 mmol H_2O, 10 mol% K_2CO_3, 3 mL solvent, 15 h; a) corrected GC yields with n-tetradecane as internal standard. b) ratio determined by GC.
* The catalyst was injected as a stock solution of $RuCl_3 \cdot 3H_2O$ (0.02 mmol) in acetone (1 mL) via syringe. The solvent was removed under vacuum before any liquid compound was added.

Other aromatic solvents such as chlorobenzene and quinoline also led to total conversions (entries 17, 18) but with lower selectivities. n-Octane on the other hand shows very good results with high yields and selectivities and therefore represents the best alternative to toluene (entry 19). Dipolar protic or aprotic solvents appear to slow down the reaction (entries 20-22). This could be explained by two effects, first, the solubility of the released KCl in

Second generation catalyst for the addition of secondary amides to alkynes

polar solvents, which could affect the efficiency of the reaction, and second, solvents with Lewis-basicity could coordinate to the ruthenium center and deactivate the active species.

Finally we found out that a temperature of 80 °C appears to be sufficient (entry 23), but a decrease in temperature under 70 °C led to low conversions (entries 24-26).

The performance of this simple system in the model reaction matches that of the expensive [Ru(cod)(met)$_2$]-derived catalyst, Moreover, ESI-MS measurements of a sample taken from a toluene solution of RuCl$_3$·3H$_2$O, n-Bu$_3$P, DMAP, K$_2$CO$_3$, water, and pyrrolidin-2-one displayed again the distinct signals corresponding to [(n-Bu$_3$P)$_2$Ru(C$_4$H$_6$NO)(C$_7$H$_{10}$N$_2$)]$^+$ and [(n-Bu$_3$P)$_3$Ru(C$_4$H$_6$NO)]$^+$ fragments (Scheme 76).

Second generation catalyst for the addition of secondary amides to alkynes

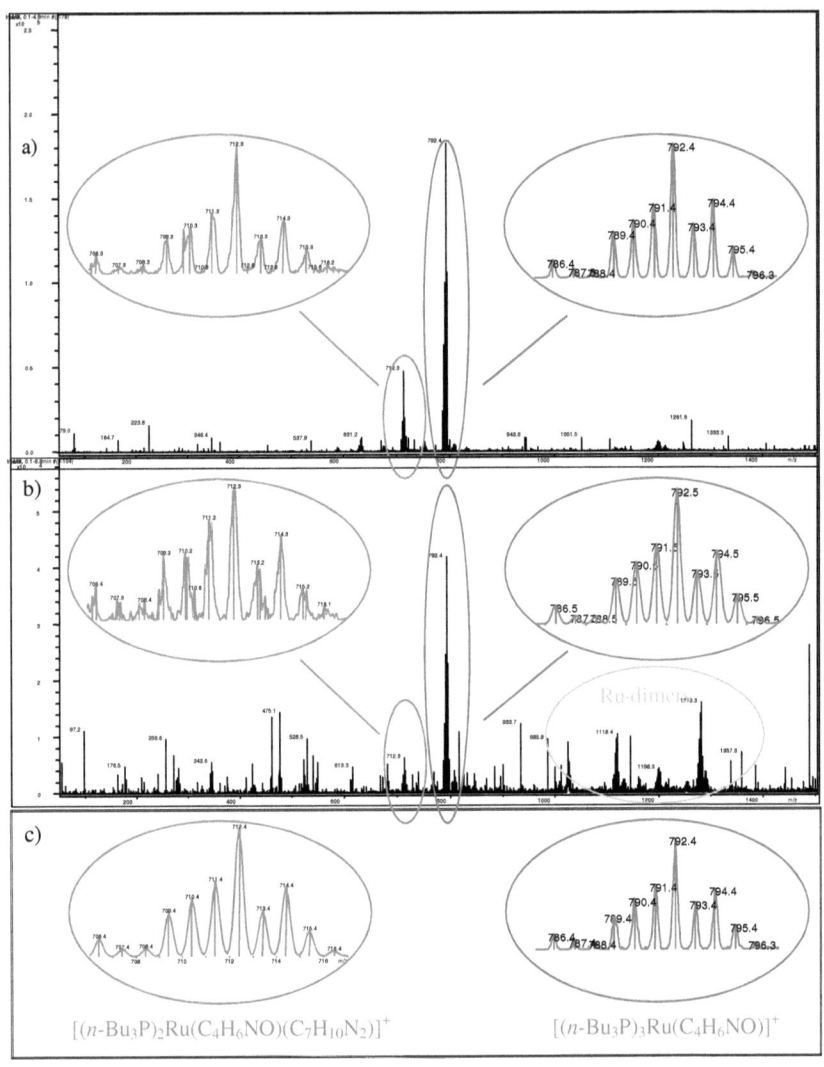

Scheme 76: ESI-MS of the catalyst systems based on a) RuCl$_3$ and b) [Ru(cod)(met)$_2$].

These results support our hypothesis that the same type of catalytically active ruthenium(II) species is generated with either method (Scheme 77).

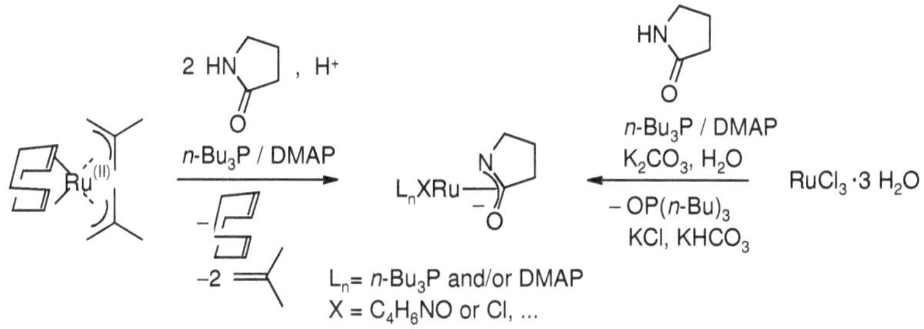

Scheme 77: catalyst preformation with [Ru(cod)(met)₂] and RuCl₃·3H₂O.

2.3.3 Scope of the reaction using the second generation catalyst

We next tested the generality of the new protocol by applying it to the synthesis of enamide derivatives from various *N*-nucleophiles and alkynes (Table 24).

Table 24: scope of the reaction.

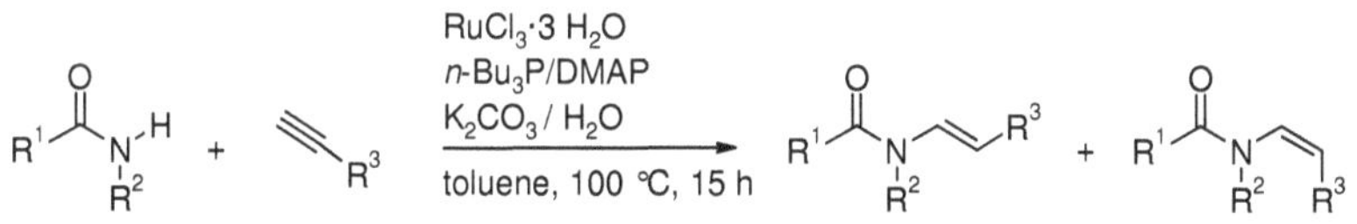

product	yield (ratio 14:15)[a]	product	yield (ratio 14:15)[a]
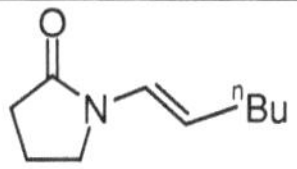 14aa	99 (32:1) (95 (25:1))[b]	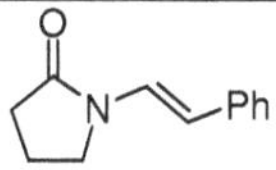 14ab	97 (22:1) (99 30:1)[b]
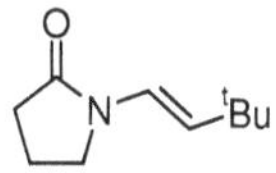 14ac	99 (40:1) (99 (30:1))[b]	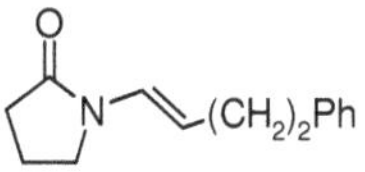 14ad	96 (25:1) (97 (30:1))[b]
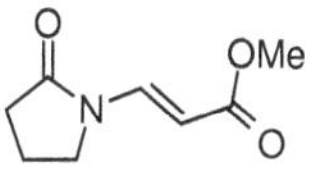 14ae	96 (40:1) (99 (30:1))[b]	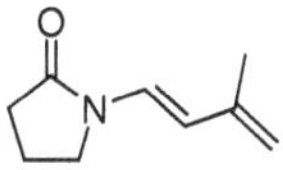 14af	97 (8:1) (98 (24:1))[b]
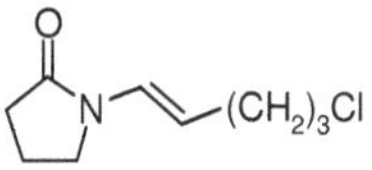	99 (33:1) (81 (30:1))[b]	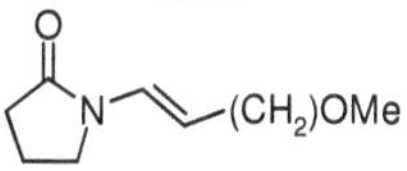	99 (24:1) (93 (8:1))[b]

- 95 -

Second generation catalyst for the addition of secondary amides to alkynes

$$R^1\underset{R^2}{\underset{|}{C(O)N}}H + \equiv\!\!-R^3 \xrightarrow[\text{toluene, 100 °C, 15 h}]{\text{RuCl}_3\cdot 3\,\text{H}_2\text{O},\; n\text{-Bu}_3\text{P/DMAP},\; \text{K}_2\text{CO}_3/\text{H}_2\text{O}} R^1\underset{R^2}{\underset{|}{C(O)N}}\!\!-\!\!\text{CH=CH}\!-\!R^3\;(14) + R^1\underset{R^2}{\underset{|}{C(O)N}}\!\!-\!\!\text{C(R}^3\text{)=CH}_2\;(15)$$

12a-p   13a-j   14   15

product	yield (ratio 14:15)[a]	product	yield (ratio 14:15)[a]
14ag (pyrrolidinone-N-CH=CH-SiMe$_3$)	90 (2:1) (69 (3:1))[b]	**14ah** (pyrrolidinone-N-CH=CH-(CH$_2$)$_8$-CH=CH$_2$)	72 (29:1) (99 (30:1))[b]
14ai (β-lactam-N-CH=CH-^{n}Bu)	35 (1:1) (70 (2:1))[b]	**14aj** (piperidinone-N-CH=CH-^{n}Bu)	91 (40:1) (94 (30:1))[b]
14ba (caprolactam-N-CH=CH-^{n}Bu)	99 (40:1) (94 (30:1))[b]	**14ca** (azocan-2-one-N-CH=CH-^{n}Bu)	30 (40:1) (86 (30:1))[b]
14da (diketopiperazine bis-N-CH=CH-^{n}Bu)	71 (40:1) (99 (30:1))[b]	**14ea** (4-acetyl-C$_6$H$_4$-N(Ac)-CH=CH-^{n}Bu)	20 (40:1) (33 (30:1))[b]
14fa (4-OEt-C$_6$H$_4$-N(Ac)-CH=CH-^{n}Bu)	93 (40:1) (99 (30:1))[b]	**14ga** (Ph-C(O)-N(Ph)-CH=CH-^{n}Bu)	20 (40:1)
14ha (H-C(O)-N(Me)-CH=CH-^{n}Bu)	35 (2.5:1) (83 (30:1))[b]	**14ia** (Me-N(Ac)-CH=CH-^{n}Bu)	10 (2.5:1) (84 (30:1))[b]
14ja (^{i}Pr-N(C(O)CH=CH$_2$)-CH=CH-^{n}Bu)	49 (40:1) (38 (30:1))[b]	**14ma** (oxazolidinone-N-CH=CH-^{n}Bu)	88 (30:1) (90 (30:1))[b]
14la			

Second generation catalyst for the addition of secondary amides to alkynes

$$R^1\text{C(O)N(H)R}^2 + \equiv\!\!-R^3 \xrightarrow[\text{toluene, 100 °C, 15 h}]{\substack{\text{RuCl}_3\cdot 3\text{ H}_2\text{O} \\ n\text{-Bu}_3\text{P/DMAP} \\ \text{K}_2\text{CO}_3/\text{ H}_2\text{O}}} R^1\text{C(O)N(R}^2)\text{CH=CHR}^3 + R^1\text{C(O)N(R}^2)\text{C(R}^3)\text{=CH}_2$$

12a-p	13a-j		14	15

product	yield (ratio 14:15)[a]	product	yield (ratio 14:15)[a]
14na (oxazolidinone with iPr, N-CH=CH-^{n}Bu)	94 (22:1) (97 (30:1))[b]	**14oa** (oxazolidinone with Ph, N-CH=CH-^{n}Bu)	94 (21:1) (84 (21:1))[b]
14pa (succinimide, N-CH=CH-^{n}Bu)	8 (1:1) (12 (1:1))[b]		

Reaction conditions: 1.00 mmol N-nucleophile,, 2.00 mmol alkyne, 3 mol% RuCl$_3$·3 H$_2$O*, 9 mol% n-Bu$_3$P, 9 mol% DMAP, 0.40 mmol H$_2$O, 15 mol% K$_2$CO$_3$, 3 mL toluene, 100 °C, 15 h; a) isolated yield, ratio determined by GC. b) isolated yields and selectivity of the [Ru(cod)(met)$_2$] based hydroamidation.
* The catalyst was injected as a stock solution of RuCl$_3$·3H$_2$O (0.02 mmol) in acetone (1 mL) via syringe. The solvent was removed under vacuum before any liquid compound was added.

As can be seen from above, linear and branched aliphatic as well as aromatic terminal alkynes bearing a range of functional groups were smoothly and selectively converted with pyrrolidin-2-one (**14aa-14aj**) in yields that were sometimes even in excess of those obtained with the first generation catalyst. Moreover, various N-nucleophiles could be added to 1-hexyne (**14ba-14pa**), including amides, lactams, bislactams and oxazolidinones.

For the products **3ba** and **3ea**, lower yields were observed. This could be explained by the competition between the substrates and the chloride ions still in solution. This effect becomes more pronounced if the N-nucleophile has difficulties to coordinate to the ruthenium center. These difficulties could be due to the ring-tension after the deprotonation of the four-membered lactam or to the important steric hindrance of the nine-membered lactam.

A similar competition was also observed for the products **14ia, 14ja, 14ka**. It is likely that these particular substrates already have difficulties to coordinate to the ruthenium-center, and then are too slow to compete with the chloride ions.

To summarize, a practical protocol for hydroamidation reactions has been discovered in which the catalyst is generated *in situ* from inexpensive ruthenium(III) chloride hydrate, tri-n-

Second generation catalyst for the addition of secondary amides to alkynes

butylphosphine or its HBF$_4$ adduct, 4-(dimethylamino)pyridine and potassium carbonate. The catalyst allows the addition of secondary amides to terminal alkynes, and yields the corresponding *E-anti*-Markovnikov enamides. The conversions observed often match or even exceed those of the costly first-generation catalyst. Moreover, the protocol could easily be scaled up to gram quantities without any losses in yield, selectivity or money.

2.4 Mechanistic studies of the hydroamidation of alkynes

To summarize, the hydroamidation of alkynes is no longer limited to the secondary amides, major changes in the protocol allow us now to add primary amides and imides to terminal alkynes in very good yields and selectivities. Moreover, great achievements were made by replacing the expensive [Ru(cod)(met)$_2$] catalyst by the more cheaper RuCl$_3$·3 H$_2$O for the secondary amides protocol.

However, the mechanism of this reaction is still unclear. A possible route to obtain this kind of regioselectivity would be the one supported by Dixneuf et al. for the addition of carboxylic acid to alkynes involving a Ru-alkylidene complexe (Scheme 78).[46]

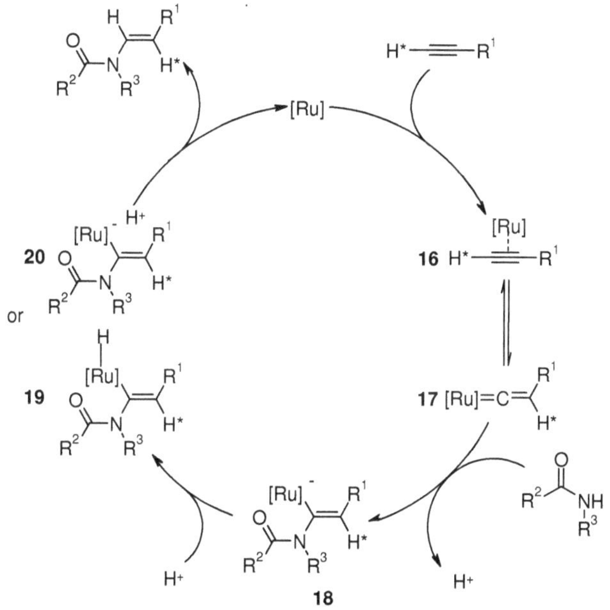

Scheme 78: mechanism involving a Ru-alkylidene.

The catalytic cycle begins by coordination of the alkyne to the metal center (**16**). Tautomerization of the complex **16** involving [1, 3] proton shift leads to the intermediate **17**. The latter undergoes a nucleophilic attack of the amide on the α–position of the ruthenium

- 99 -

centre to give intermediate **18**. Protonation of **18** followed by a reductive elimination delivers the desired product and recreates the original ruthenium center.

Another plausible mechanism would be the one developed by Wakatsuki et al. for the addition of water to alkynes. Scheme 79 depicts this mechanism adapted to our reaction.[36]

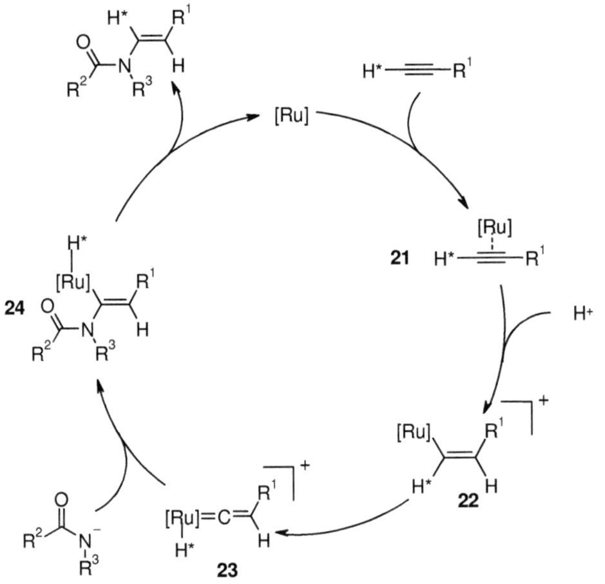

Scheme 79: mechanism according to Wakatsuki et al.

This mechanism begins again with coordination of the alkyne to the metal center to form the intermediate **21**. External attack of the latter by H^+ leads to a vinylic–ruthenium species **22**, which is stabilized by a [1, 2]-shift of hydrogen to the metal. Attack of a deprotonated amide to the vinylidene-ruthenium complex **23** leads to the intermediate **24**, which reductively eliminates the enamide with release of the original ruthenium complex.

Finally, we also developed a mechanism based on our observations. Our approach differs from the previous ones with regards to the coordination step. From our point of view, the ruthenium center first coordinates to the amide moiety (Scheme 80).

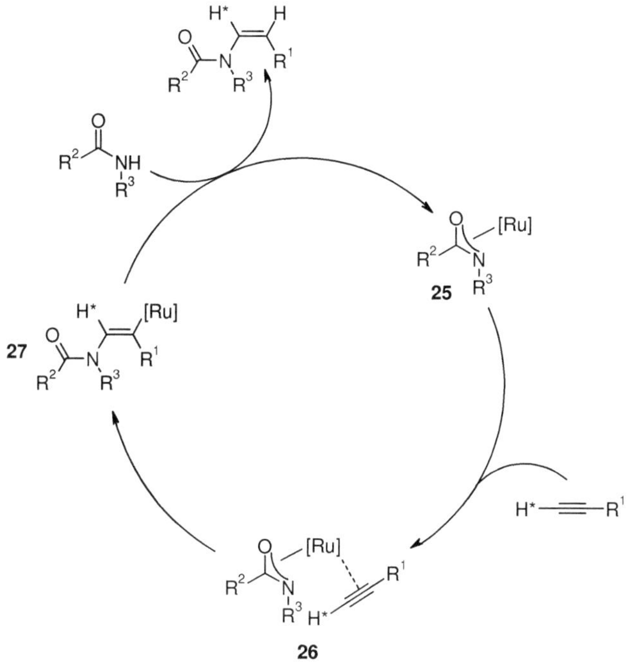

Mechanistic studies of the hydroamidation of alkynes

Scheme 80: mechanism involving activation of the amide.

Our proposed catalytic cycle starts from a ruthenium-amide complex (**25**). Initially, the alkyne adds to the Ru-center, possibly displacing one of the ligands under formation of complex **26**. The amide anion then attacks the coordinated alkyne, and the enamide product is finally released from the intermediate **27** by protonation, thus regenerating the original catalytic active species **25**.

In order to pinpoint which of these mechanisms could be best applied to our reaction, we decided to perform the reaction with deuterated alkynes, so that we could observe whether or not the terminal proton is shifted.

We performed the addition of pyrrolidin-2-one, benzamide, and succinimide to deuterated 1-hexyne (Scheme 81, respectively equation 1, 2 and 3) with the best conditions developed for each of the substrates.

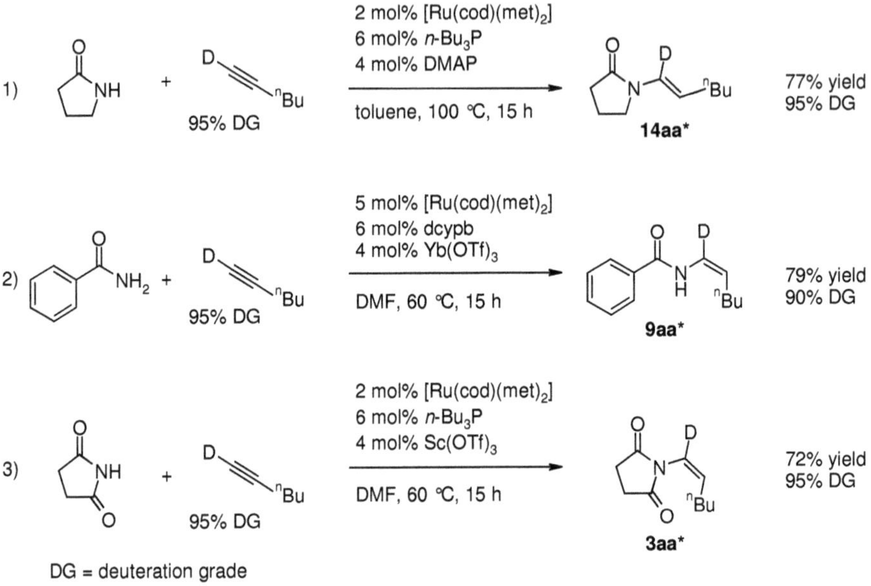

Scheme 81: deuteration experiments.

1) Pyrrolidinone-NH + D-C≡C-^{n}Bu (95% DG) → 14aa* (77% yield, 95% DG)
   Conditions: 2 mol% [Ru(cod)(met)$_2$], 6 mol% n-Bu$_3$P, 4 mol% DMAP, toluene, 100 °C, 15 h

2) Benzamide-NH$_2$ + D-C≡C-^{n}Bu (95% DG) → 9aa* (79% yield, 90% DG)
   Conditions: 5 mol% [Ru(cod)(met)$_2$], 6 mol% dcypb, 4 mol% Yb(OTf)$_3$, DMF, 60 °C, 15 h

3) Succinimide-NH + D-C≡C-^{n}Bu (95% DG) → 3aa* (72% yield, 95% DG)
   Conditions: 2 mol% [Ru(cod)(met)$_2$], 6 mol% n-Bu$_3$P, 4 mol% Sc(OTf)$_3$, DMF, 60 °C, 15 h

DG = deuteration grade

As we solely detected the formation of products with the deuterium remaining in the 1-position for all of these reactions, reaction pathways via 1,3 proton shift have to be excluded for these particular transformations.

Only the two other reaction mechanisms remain plausible, the one proposed by Wakatsuki et al. and the one that we proposed. Unfortunately, at this stage of the mechanistic studies, it is not possible to decide if one of these mechanisms is the right one or not.

A decisive point would be the isolation of reaction intermediates, however this was not possible so far. Our attempts to preform the supposedly active catalytic species and to isolate it failed because of the too high sensitivity of ruthenium complexes.

But in view of the results obtained with the ESI-MS experiments (see previous chapter), we are able to propose an hypothetic mechanism, which from our point of view describes best what happens in the reaction vessels (Scheme 82). This mechanism is the one we propose for the addition of primary amides to alkynes.

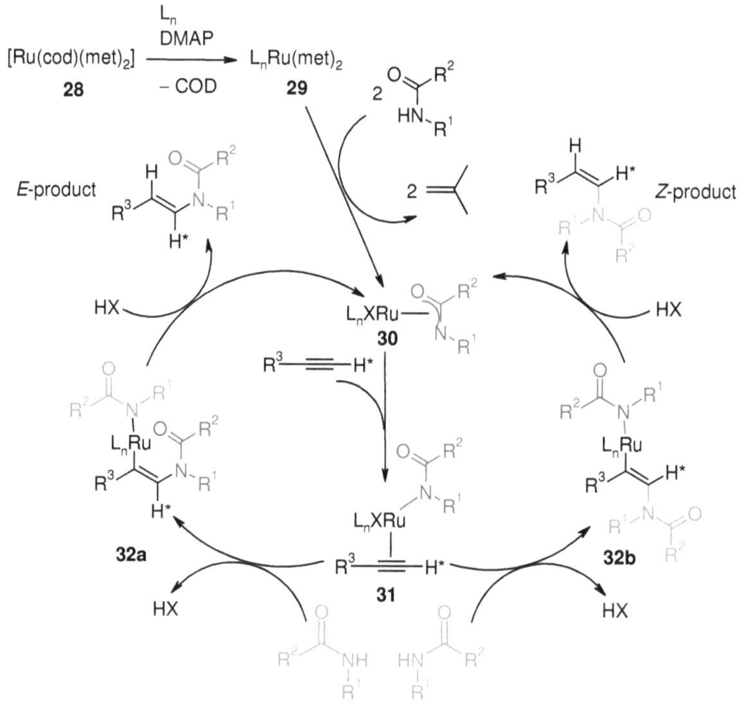

Mechanistic studies of the hydroamidation of alkynes

L_n = phosphine and/or co-ligand (i.e. DMAP)
X = $NR^1C(O)R^2$ (or OTf in the case of the primary amide protocol)

Scheme 82: proposed catalytic cycle for the addition of secondary amides to alkynes.

Supported by the ESI-MS experiments, we are sure that none of the original ligands of the elaborate Ru-precursor **28** actually remains bound to the metal center, so that our catalytic cycle starts from L_nXRu-amide complex **30**. Coordination of the alkyne to the ruthenium center leads to the complex **31**. Attack of the amide anion to the coordinated alkyne forms the ruthenium complexes **32a** and **32b**. Depending on whether the coordinated amides attacks the alkyne and is replaced by an external amide anion or whether the latter directly attacks the coordinated alkyne, the (Z)-configured (**32a**) or (E)-configured (**32b**) ruthenium amide complex forms. Protonolysis of the intermediate would probably proceed with preservation of the stereochemistry, regenerating the initial ruthenium species (**30**) and completing the catalytic cycle.

To conclude, we were able to propose a mechanism supported by our experimental observations. However, it is strictly hypothetic and the reaction pathway postulated by Wakatsuki et al. could not be excluded.

2.5 Addition of thioamides to alkynes

2.5.1 Problem statement

The particular reliability of the methodology developed for the addition of amides to alkynes encouraged us to extend this strategy to the rapid synthesis of thioenamides. As seen before, enamides are well described in the literature, however, their sulfurated analogues which could be of considerable interest as a structural variant of the enamide subunit are only scarcely represented. At present, treatment of the analogous enamides by Lawesson's[88] reagent or other similarly aggressive sulfurizing[89] agents is the only possibility to obtain the thioenamide structure. But it is evident that the most sensitive functional groups would not be tolerated by such processes. Thus, an easy and concise method for the synthesis of thioenamides is probably the key to the development of its chemistry. As for the synthesis of enamides, an expedient method for the preparation of thioenamides would be the catalytic addition of thioamides to alkynes (Scheme 83).[90]

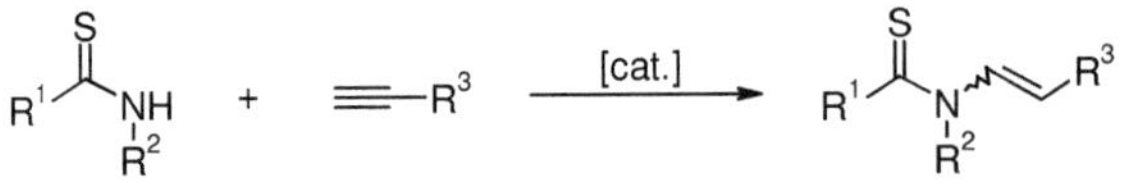

Scheme 83: addition of thioamides to alkynes.

However, in comparison to amides, thioamides present a more pronounced ambident nucleophilic nature due to the softness of the sulfur and nitrogen atoms, and thus can react on the nitrogen or sulfur terminus depending on the electrophile used, following the HSAB principle (Scheme 84).[91]

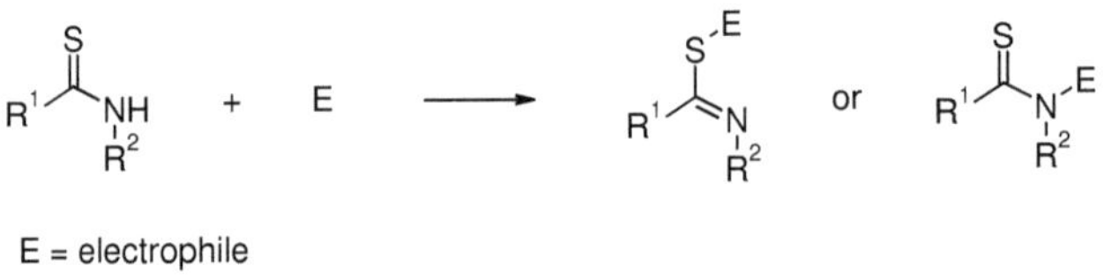

E = electrophile

Scheme 84: addition of thioamides to an electrophile.

This additional chemoselectivity issue, in addition to the regio- and stereocontrol, also had to be controlled in our planned transition metal-catalyzed reaction, along with the fact that sulfur-containing compounds are known catalyst poisons due to their strong interaction with late transition metals. This project was pursued together with Dipl.-Chem. Kifah M. Salih and is part of both our Ph.D. theses.

2.5.2 Development of a new catalyst system

We chose thiopyrolidinone and 1-hexyne as test substrates for our studies because of their good availability (Table 25).

Table 25: test of different additives.

entry	additive	yield (%)[a]	ratio (35aa:36aa)[b]
1	DMAP	40	6:1
2	K_2CO_3	84	5:1
3	Na_2CO_3	88	5:1
4	LiCl	87	7:1
5	CsCl	93	6:1
6	3 Å MS	93	11:1

Reaction conditions: 0.50 mmol thiopyrolidinone, 1.00 mmol 1-hexyne, 2 mol% [Ru(cod)(met)$_2$], 6 mol% n-Bu$_3$P, 4 mol% additive or 250 mg 3 Å molecular sieves, toluene (1.5 mL), 100 °C, 15 h. a) corrected GC yields with n-tetradecane as internal standard. b) ratio determined by GC.

Our first attempt using the catalytic system developed for the addition of secondary amides to alkynes did not reach the expected level of activity and N-((E)-hex-1-enyl)pyrrolidine-2-thione was obtained rather in low yields and unsatisfactory selectivity (entry 1). This may be attributed to the substantially higher acidity of thioamides compared to amides (pK_a of 2-pyrrolidone, 24.2; pyrrolidine-2-thione, 18.1).[92] The identity of the product obtained was confirmed by comparison to a sample that was prepared *via* a literature procedure using the analogous enamides as starting material.[91b]

Addition of thioamides to alkynes

Higher yields were obtained by substituting DMAP with inorganic bases, although without improvement in terms of selectivity. This enhancement in yield is probably caused by the effect of the counter ions of the inorganic base (entries 2, 3). Mild Lewis acids (entries 4-6) such as lithium chloride, cesium chloride or molecular sieves improved both yield and selectivity, with the added beneficial effect of the latter being capable to remove the moisture contained in the reaction mixture, thus preserving the product from an hydrolysis.

Test of different ruthenium sources

Among all the ruthenium sources tested, only [Ru(cod)(met)$_2$] was particularly effective. Other ruthenium catalysts were also active but they never reached its level of activity and selectivity. (Table 26)

Table 26: test of different ruthenium sources.

entry	[Ru-cat.]	yield (%)[a]	ratio (35aa:36aa)[b]
6	[Ru(cod)(met)$_2$]	93	11:1
7	[(p-cymene)RuCl$_2$]	72	10:1
8	[Ru(cod)Cl$_2$]	80	9:1
9	[Ru$_3$(CO)$_{12}$]	0	nd
10	[RuCl$_3$]	33	8:1

Reaction conditions: 0.50 mmol thiopyrrolidinone, 1.00 mmol 1-hexyne, 2 mol% [Ru-cat.], 6 mol% n-Bu$_3$P, 250 mg 3 Å molecular sieves, toluene (1.5 mL), 100 °C, 15 h. a) corrected GC yields with n-tetradecane as internal standard. b) ratio determined by GC.

Test of different ligands

To achieve the completion of the reaction and make the reaction more selective, we then tested some other ligands.

Table 27: test of different ligands.

entry	ligand	yield (%)[a]	ratio (35aa:36aa)[b]
6	n-Bu$_3$P	93	11:1
11	P(2-Fur)$_3$	66	>20:1
12	PPh$_3$	43	5:1
13	t-Bu$_3$P	10	nd
14	n-Oct$_3$P	96	16:1
15	dppm	80	4:1
16	dcypm	15	1:1

Reaction conditions: 0.50 mmol thiopyrrolidinone, 1.00 mmol 1-hexyne, 2 mol% [Ru(cod)(met)$_2$], 6 mol% ligand (or 1.5 mol% for chelating ligand), 250 mg 3 Å molecular sieves, toluene (1.5 mL), 100 °C, 15 h. a) corrected GC yields with n-tetradecane as internal standard. b) ratio determined by GC.

As experienced during our previous reaction optimizations, electron-poor and sterically hindered phosphines slow down the reaction (entries 11-13). Chelating phosphines did not lead to any improvement, probably because of their too strong coordination to the metal center, and in the case of bis(dicyclohexyl-phosphino)methane, the yield of the reaction dramatically dropped to 15% (entries 15, 16). As in the addition of secondary amides to alkynes, only tri-alkyl phosphines led to the desired product with high yields and selectivities. A small improvement compared to the previous protocol was made by using tri(n-octyl)phosphine (entry 14), instead of tri(n-butyl)phosphine. It is likely that the stronger steric effect of the former is responsible for this enhancement of the selectivity.

Optimization of the reaction conditions

We finally tested the reaction in other solvents in order to find out if toluene could be substituted by another solvent with positive effect on the selectivity of the transformation. Important was also to observe if the reaction temperature could be reduced without loss of activity (Table 28).

Table 28: optimization of the reaction conditions.

entry	solvent	temp. (°C)	yield (%)[a]	ratio (35aa:36aa)[b]
14	toluene	100	96	16:1
17	glyme	"	85	12:1
18	EtOH	"	55	10:1
19	DMF	"	98	6:1
20	mesitylene	"	91	13:1
21	"	80	82	12:1
22	"	120	87	16:1

Reaction conditions: 0.50 mmol thiopyrrolidinone, 1.00 mmol 1-hexyne, 2 mol% Ru-source, 6 mol% n-Bu$_3$P, 250 mg 3 Å molecular sieves, solvent (1.5 mL), 15 h. a) corrected GC yields with n-tetradecane as internal standard. b) ratio determined by GC.

Nonpolar solvents were most effective, as expected reaction performed in mesitylene leads to almost the same results as in toluene. However, other solvents can also be used, although with either lower yields in the case of ethanol, or loss of selectivity if DMF is used (entry 19). Finally mesitylene was used for testing different temperatures because of its higher boiling point. It appears that 100 °C represents the best compromise between turnover rate and stereoselectivity. Performing the reaction at lower or higher temperature slowed down the reaction (entries 20-22).

Addition of thioamides to alkynes

Under these optimized conditions (entry 14), we exclusively observed the formation of the (E)-anti-Markovnikov product, and the hypothetic addition occurring from the S-rather than the N-terminus of the thioamide could never be detected.

2.5.3 Scope of the reaction

To investigate the scope of this new hydroamidation protocol, we applied it to the addition of a broad variety of nitrogen nucleophiles to several terminal alkynes.

Table 29: scope of the reaction.

product	yield (%)^a (ratio 35:36)^b	product	yield (%)^a (ratio 35:36)
35aa	98 (16:1)	35ab	97 (13:1)
35ac	95 (7:1)	35ad	79 (15:1)^c
35ae	76 (7:1)	35af	76 (11:1)^c
35ag	57 (19:1)	35ah	88 (1:1)^c

Addition of thioamides to alkynes

$$R^1\underset{R^2}{\overset{S}{\underset{|}{C}}}NH + \overset{}{\equiv}R^3 \xrightarrow[\text{toluene, 100 °C, 15 h}]{[\text{Ru(cod)(met)}_2]\ n\text{-Oct}_3P\ 3\text{ Å MS}} R^1\underset{R^2}{\overset{S}{\underset{|}{C}}}N\diagup\!\!\diagdown R^3 + R^1\underset{R^2}{\overset{S}{\underset{|}{C}}}N\diagup\!\!\diagdown R^3$$

33   34   35   36

product	yield (%)[a] (ratio 35:36)[b]	product	yield (%)[a] (ratio 35:36)
35ai	71 (3:1)[c]	35aj	97 (6:1)
35bj	89 (5:1)[c,d]	35cj	84 (30:1)[c]
35dj	98 (30:1)[c]	35ej	97 (30:1)
35fj	91 (15:1)[c]	35gj	42 (2:1)[c]

Reaction conditions: 1.00 mmol N-nucleophile, 2.00 mmol alkyne, 2 mol% [Ru(cod)(met)$_2$], 6 mol% n-Oct$_3$P, 500 mg 3 Å molecular sieves, toluene (3 mL), 15 h. a) isolated yield. b) ratio determined by GC. c) 5 mol% [Ru(cod)(met)$_2$], 15 mol% n-Oct$_3$P. d) ratio determined by ^{1}H NMR.

On the one hand, pyrrolidine-2-thione added smoothly to various alkynes, among them alkyl- and aryl-substituted alkynes, trimethylsilylacetylene, and conjugated enynes. Surprisingly, low yields were observed in the case of trimethyl-prop-2-ynyloxy-silane (**35ag**). Since no dimerization products were observed with this alkyne, we assumed that it decomposed during the reaction because of its sensitivity.

On the other hand, phenylacetylene reacted with a range of secondary thioamides bearing aromatic or aliphatic substitutents, as well as with a thiocarbamate. In most cases, the products were obtained in good yields and high selectivities for the (*E*)-configured *anti*-

Markovnikov-thioenamides. We still faced a problem by using [1,3]oxazinane-2-thione as it seems that the dimerization reaction of phenylacetylene is faster than the addition reaction, thus leading to low yields (**35gj**).

2.5.4 Reversal of the selectivity

In an attempt to invert the stereoselectivity of the addition, we combined the ligand with the lowest *E*-selectivity, bis(dicyclohexylphosphino)methane (dcypm), with various additives, and found that the *Z*-isomer can indeed be obtained as the major product with 76% yield and 1:2 selectivity in the presence of potassium *tert*-butoxide.

Table 30: scope of the *Z*-selective protocol.

product	yield (%)[a] (ratio 35:36)[b]	product	yield (%)[a] (ratio 35:36)
36aa	74 (1:2)	36ab	70 (1:8)
36ac	64 (1:2)		

Reaction conditions: 1.00 mmol *N*-nucleophile, 2.00 mmol alkyne, 2 mol% Ru(cod)(met)$_2$, 3 mol% dcypm, 4 mol% KO^{t}Bu, toluene (3 mL), 15 h. a) isolated yield. b) ratio determined by GC.

The scope of the complementary (*Z*)-selective hydrothioamidation was also tested on a number of examples. This protocol appears to be as high-yielding, although the level of stereoselectivity never exceeded a moderate 1:8 ratio. Further work is aimed at extending the synthetic utility of this complementary protocol by improving the (*Z*)-selectivity with the help of customized ligands.

2.6 Markovnikov addition of secondary amides to alkynes

Finally, we concentrated our efforts in finding a protocol for the Markovnikov addition of amides to alkynes. Products resulting from such a transformation are of particular importance from a synthetic point of view (i.e. enantioselective hydrogenation, etc...), they are also often represented in the synthesis of natural products. Following the methodology developed for the synthesis of the *anti*-Markovnikov enamides, we tried to identify a catalytic system active only for the Markovnikov addition of amides to alkynes.

The experimental results obtained so far show the complexity of our Ru-catalyzed reaction protocols. It was very disappointing that no Markovnikov product had formed under any set of conditions, and solely from the proposed catalytic cycle, it is hard to rationalize this decisive difference of reactivity between amides and carboxylic acids. This is why we decided to seek inspiration from related literature procedures on the hydroamination of alkynes and hydroamidation of alkenes.

2.6.1 Based on the Markovnikov selective hydroamination of alkynes of Uchimaru[67]

Uchimaru presented an interesting method for the addition of secondary amines to terminal alkynes catalyzed by [Ru$_3$CO$_{12}$] in the absence of solvent (Scheme 85).

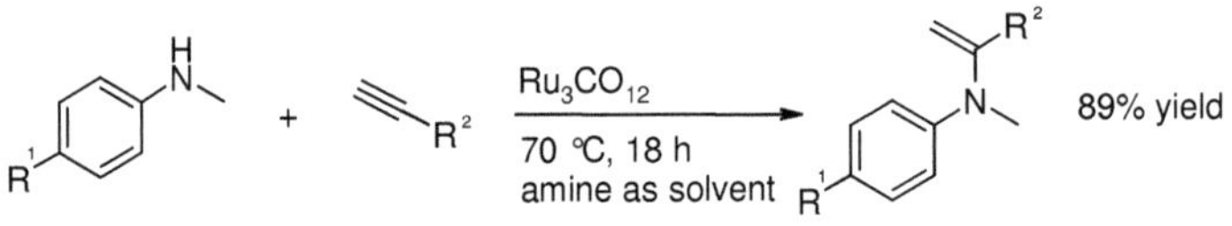

Scheme 85: addition of secondary amines to terminal alkynes according to the method of Uchimaru.

The proposed mechanism of the reaction consists of an oxidative addition of the amine N–H bond to a ruthenium center giving the amidoruthenium hydride, a coordination of the alkyne to the metal, followed by an intramolecular attack of the nitrogen to the C–C triple bond and a reductive elimination of the enamine.

The yields were moderate and the scope limited to very few combinations of starting materials, but we still decided to investigate whether this reaction might serve as a starting point for further investigations.

Markovnikov addition of secondary amides to alkynes

Table 31: screening table for the hydroamination of alkynes.

$$R^1R^2NH + \text{HC≡C-Ph} \xrightarrow[\text{solvent, 70 °C, 15 h}]{\text{[cat.] ligand additive}} R^1R^2N\text{-C(=CH}_2\text{)-Ph}$$

       37   38   39

entry	N-nucleophile	[cat.] / ligand	solvent	yield (%)[a]
1	PhNH(Me)	[Ru$_3$CO$_{12}$]	neat	25
2	"	"	toluene	13
3	"	"	DMF	14
4	"	[Ru(cod)(met)$_2$] / n-Bu$_3$P	neat	0
5	4-HO-C$_6$H$_4$-NH-C(O)CH$_3$	Ru$_3$CO$_{12}$	neat	0
6	2-pyrrolidinone	"	neat	0
7	"	"	DMF	0
8	"	"	toluene	0
9	succinimide	"	neat	0
10	"	"	DMF	0
11	"	"	toluene	0

Reaction conditions: 1.00 mmol of amine / amide, 2.00 mmol of phenylacetylene, 2 mol% Ru precursor, 6 mol% ligand, 4 mol% additive, 3 ml solvent, 70 °C, 15 h. a) corrected GC yields with n-tetradecane as internal standard.

Unfortunately, our own experience with this reaction was not at all encouraging. All of our attempts to reproduce the results of Uchimaru failed, and 25% was the best yield obtained, instead of the 89% reported. Other catalyst systems also gave low yields (Entries 1-4). Applying this system to secondary amides and imides did not result in any conversion, so that we abandoned this approach.

2.6.2   Based on the hydroamidation of alkyl propiolates from Trost[75]

An interesting method for the synthesis of dehydroamino acids has been described by Trost, who added N–H nucleophiles to alkynoates under formation of 2-N–H substituted acrylates. Phosphines as catalysts induce isomerization of the alkynoates to 2,4-dienoates, which then undergo addition of pronucleophiles to the 4-position (Scheme 86). He optimized a system capable of adding imides to some terminal alkynes to give the Markovnikov product, just by using triphenylphosphine, sodium acetate and acetic acid.

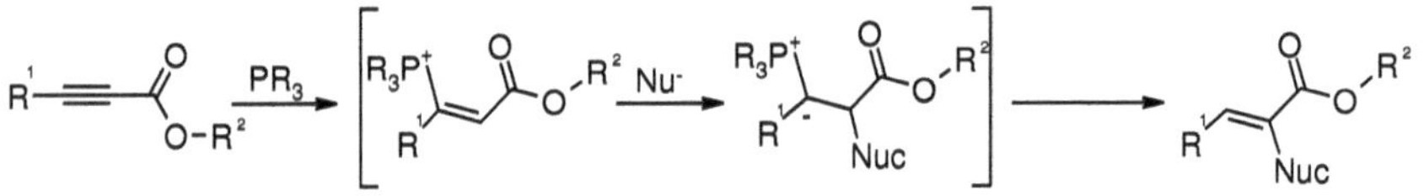

Scheme 86: Alkynoates-nucleophilic addition at the α-position.

We chose the reaction of phthalimide with ethyl propiolate as a model system to verify these results, and were pleased to find that we could reproduce the 98% yield described in the literature (Table 32, entry 1).

Markovnikov addition of secondary amides to alkynes

Table 32: addition of phthalimide to alkynes.

[reaction scheme: phthalimide (40) + alkyne (41) with [cat.], ligand, additive in toluene, 105 °C, 15 h → N-vinyl phthalimide (42)]

entry	alkyne	[cat.]	ligand	additive	yield (%)[a]
1	propargyl ester (CH≡C-C(O)-O-Et)	-	PPh$_3$	NaOAc, AcOH	98
2	"	-	"	"	97
3	hex-1-yne	-	"	"	0
4	phenylacetylene	-	"	"	0
5	propargyl methyl ether	-	"	"	0
6	propargyl ester	-	t-Bu$_3$P	-	48
7	"	-	PCy$_3$	-	61
8	"	-	n-Bu$_3$P	-	34
9	"	[Ru(cod)(met)$_2$]	PPh$_3$	-	38
10	hex-1-yne	"	"	-	0

Reaction conditions: 1.00 mmol phthalimide, 2.00 mmol of alkyne, 2 mol% [cat.], 6 mol% ligand, 4 mol% additive, 3 ml solvent, 105 °C, 15 h. a) corrected GC yields with n-tetradecane as internal standard.

This system is also efficient for the addition of other imides, e.g. glutarimide, phthalimide and succinimide to propiolates, although with very low yields. However, it was not possible to extend this method to other alkynes. Indeed, the alkynes described in the publication were the only ones able to react with the imides (entries 2-5).

We performed a series of experiments in order to further optimize the catalyst system (entries 6-9), and possibly extend its scope with regard to the alkynes. However, it turned out that even small change in the phosphine or additives, or the use of a coordinating metal all slowed down the reaction. As expected from the mechanism, we never observed any conversion of imides with non-activated alkynes such as 1-hexyne; the reaction is strictly limited to conjugated alkynes.

2.6.3 Following up on an unknown compound observed in our hydroamidations

We monitored all of our reactions by GC and GC/MS using *n*-tetradecane as the internal standard. Usually, we observed a maximum of two peaks with the mass of the desired product, which always turned out to be the two possible *anti*-Markovnikov stereoisomers. Due to their similar polarity, the separation of such a pair of isomers by column chromatography was usually quite challenging, especially when one of them was formed in small amounts.

We were very excited to see that in some samples, a third peak with the mass of the desired product was observed, leading us to conclude that this must be the long awaited Markovnikov product (Figure 2) However, because it was formed in rather low yields and had almost the same polarity as one of the other isomers, its isolation proved to be extremely difficult and time-consuming.

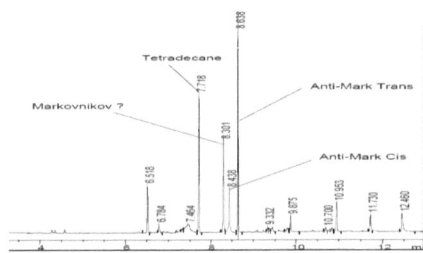

Figure 2: GC control spectra of the reaction.

We thus decided to optimize the reaction parameters to increase the selectivity for what we believed to be the Markovnikov product. We chose the reaction of pyrrolidinone with 1-hexyne as a model system and investigated the activity of several ruthenium catalysts under different conditions (Table 33).

Table 33: addition of pyrrolidinone to 1-hexyne.

entry	[cat.]	T °C	ligand	additive	yields (%)[a] (45aa:46-47aa)
1	[Ru(cod)(met)$_2$]	100	n-Oct$_3$P	Sc(OTf)$_3$	58 (1:4)
2	AuI	"	"	"	22 (>1:50)
3	[Ir(coe)Cl)]$_2$	"	"	"	0 (nd)
4	[Rh(cod)(acac)]	"	"	"	63 (1:4)
5	[Ir(cod)(acac)]	"	"	"	50 (1:12)
6	AuCl$_3$	"	"	"	0 (nd)
7	Rh(nbd) BF$_4$	"	"	"	75 (1:3.5)
8	-	"	"	"	0 (nd)
9	[Ru(cod)(met)$_2$]	50	"	"	72 (>1:50)
10	"	150	"	"	59 (1:3)
11	"	200	"	"	0 (nd)
12	"	150	n-Oct$_3$P	Zn(OTf)$_2$	51 (1:4)
13	"	"	"	Li(OTf)	28 (>1:50)
14	"	"	"	Al(OTf)$_3$	49 (1:4)
15	"	150	PPh$_3$	Zn(OTf)$_2$	10 (>1:50)
16	"	"	i-Pr$_3$P	"	8 (1:1)
17	"	"	P(2-Fur)$_3$	"	0
18	"	150	dcpe	Zn(OTf)$_2$	60 (1:1)

Reaction conditions: 1.00 mmol of pyrrolidinone, 2.00 mmol of 1-hexyne, 2 mol% of [cat.], 6 mol% ligand, 4 mol% additive, 15 h; a) corrected GC yields with n-tetradecane as internal standard and ratio determined by GC.

The first time we observed the formation of a "new product" (**45aa**) was when using [Ru(cod)(met)$_2$] / n-Oct$_3$P in DMF at 100 °C in the presence of scandium triflate (entry 1).

Different combinations of Ru-sources with n-Oct$_3$P used instead of Rh(nbd)BF$_4$ and [Ru(cod)(met)$_2$] (Entries 1, 7) did not give any conversion to **45aa** (Entries 2-8).

Based on the assumption that the Markovnikov product is a thermodynamic product, we also tested the reaction at different temperatures. At low temperatures, **14-15aa** was obtained as the major product in moderate yield (Entry 9). Only when we performed the reaction at higher temperatures, we observed an improvement of the selectivity to the desired new product (Entry 10). Nevertheless, the temperature was limited to a maximum of 200 °C due to a thermal degradation of the starting materials (Entry 11).

The test of some additives revealed that organic bases such as DMAP were again ineffective, and only specific triflates were able to give **45aa** albeit in very low yields (Entries 12-14). The use of electron-poor phosphines (Entry 15) did not give any product, and only ligands such as n-Oct$_3$P and dcpe (Entry 18) allowed us to obtain enough of **45aa** to permit us the isolation of the desired compound.

Unfortunately, NMR spectroscopy revealed unambiguously that this compound was not the desired Markovnikov, but rather a double-bond isomer of the *anti*-Markovnikov product (Figure 3), which may have formed *via* an allyl-ruthenium intermediate.

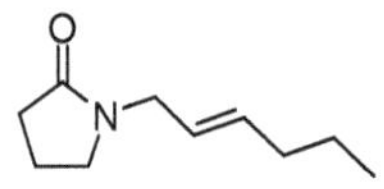

45aa

Figure 3: product resulting from an isomerisation of the double bond.

2.6.4 Using a phosphine as catalyst

Our last attempt to reverse the regioselectivity in favor of the Markovnikov product consisted in following a protocol described by Inanaga et al. where they described the *anti*-Markonikov addition of alcohol to methyl propiolate catalyzed by n-Bu$_3$P (Scheme 87).[93]

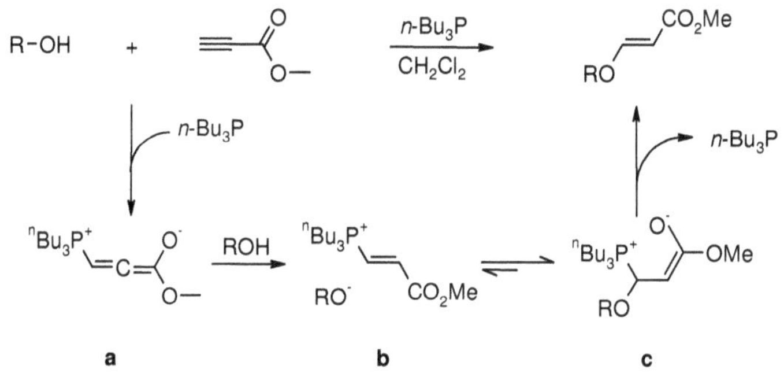

Scheme 87: addition of alcohol to alkynes by Inanaga et al.

They assumed that n-Bu$_3$P attacks the β-carbon of methyl propiolate leading to the phosphonium enolate **a**. It then abstract protons from alcohol to give the intermediate **b**. Consecutive addition of the alkoxy anion to the β-trialkylphosphonium cation and elimination of the phosphine generates the desired product.

By reproducing the same experiment using toluene as solvent for the addition of pyrrolidinone to phenylacetylene, we were pleased to find out that only one product resulting from the Markovnikov addition was obtained (Scheme 88). The identity of the product isolated was confirmed by comparison of its ^{1}H-NMR spectra with a literature spectrum.

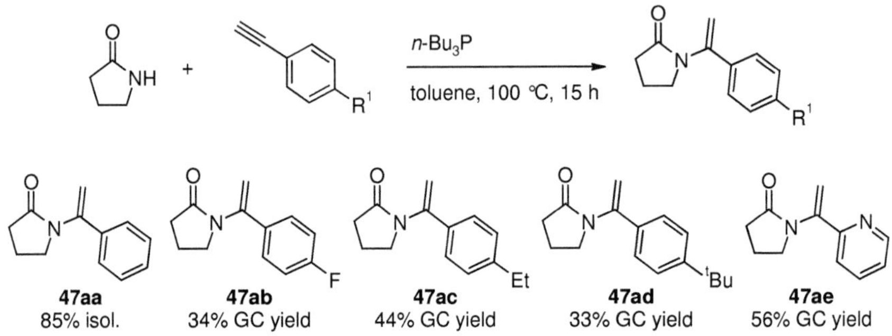

Markovnikov addition of secondary amides to alkynes

Scheme 88: Markovnikov addition of amides to alkynes.

- **47aa** 85% isol.
- **47ab** 34% GC yield
- **47ac** 44% GC yield
- **47ad** 33% GC yield
- **47ae** 56% GC yield

However, this methodology only allows the addition of pyrrolidinone to phenylacetylene derivatives. Attempts to reproduce this reaction using other amides or alkynes only led to the recovery of the starting materials. But to the best of our knowledge it represents the first and only Markovnikov addition of amides to alkynes, and this reaction is under intensive studies in our group. The mechanism of the reaction is also unclear and mechanistic investigations are currently ongoing.

3 Summary and future aspects

Progress has been made in the search for ideal hydroamidation of alkynes.

In summary, the main achievements include:

- The development of a new catalytic system for the addition of imides to alkynes. The use of Yb(OTf)$_3$ as an additive enhanced the nucleophilicity of the imides, thus allowing us the selective Z-*anti*-Markovnikov hydroamidation of terminal alkynes. Moreover, the stereoselectivity of the reaction could be controlled by changing the ligand employed, and a protocol was also developed for the exclusive syntheses of *E*-enimides.

- The development of a protocol for the addition of the primary amides to alkynes. We were able to control the rate of the reaction so that only the product resulting from the addition of amide to a single equivalent of alkyne was obtained. The reaction proceeded smoothly delivering the thermodynamically unfavored Z-enamide in high yields and with high selectivities. The *E*-isomer was obtained by a simple isomerization reaction.

- The replacement of the expensive [Ru(cod)(met)$_2$] by the more readily available [RuCl$_3$•3 H$_2$O], which proves to be as effective for this reaction. Moreover, the use of the air stable [*n*-Bu$_3$HP]$^+$[BF$_4$]$^-$ as a ligand was found to be also effective making this reaction advantageous, especially for large-scale applications.

- The development of a methodology for the addition of a new class of *N*-nucleophiles to alkynes, namely the thioamides. No addition occurring through the sulphur atom was noticed and only thioenamide products were observed with high *E*-selectivities.

Summary and future aspects

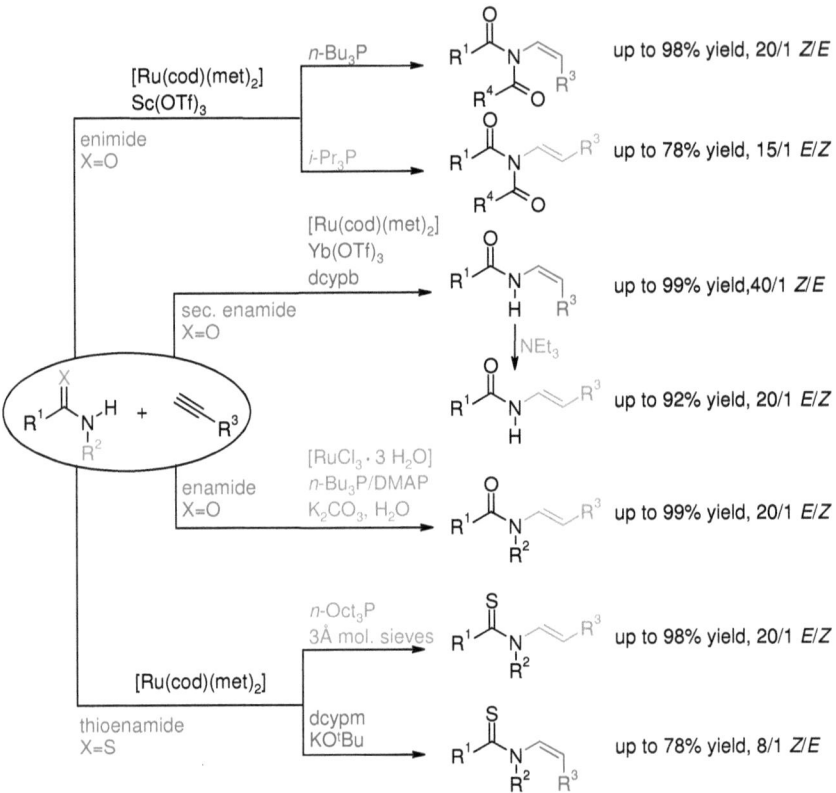

Scheme 89: addition of amides to alkynes.

In our ongoing research, we plan to extend the scope of this reaction by using new substrates. One direction is to use other *N*-nucleophiles such as sulfonamides, or more complex and sensitive amides. Another research avenue will focus on using other alkynes, for example, the selective addition of amides to disubstituted alkynes. We also plan to apply this methodology to the synthesis of some natural product (Scheme 90).

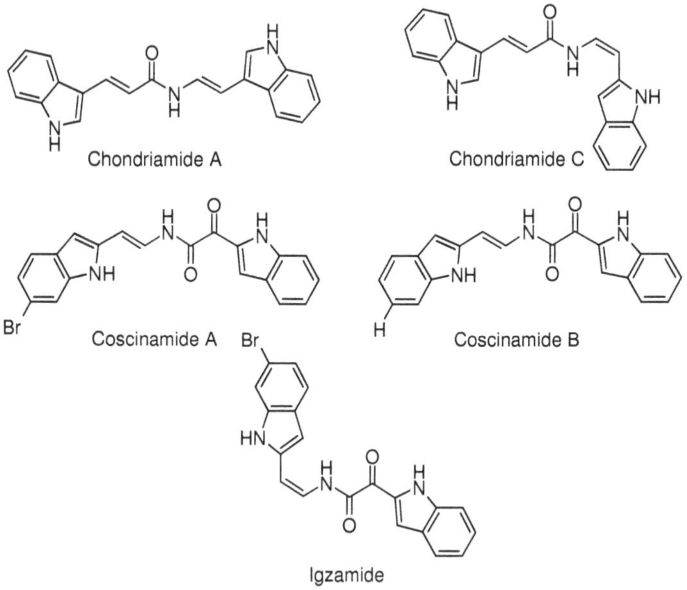

Summary and future aspects

Scheme 90: possible natural products accessible *via* hydroamidation of Alkynes.

Mechanistic studies including *in situ* ESI-MS and NMR showed the formation of a ruthenium(II)phosphinamide-complex for the addition of secondary amides to alkynes. Based on these results and the experiments performed using deuterated starting materials, we were able to propose a new reaction pathway for the hydroamidation.

Nevertheless, only the isolation and characterization of the intermediates will permit us to fully understand this reaction, but so far the isolation of these highly sensitive complexes failed.

Finally we disclosed a first protocol for the Markovnikov addition of amides to alkynes. Even if this reaction only allows the addition of pyrrolidinone to phenylacetylene analogous alkynes, for the moment it represents the proof that such a reaction is possible and also a very exciting starting point for further optimizations.

We are persuaded that the catalytic system can be optimized fully and finally employed to a broad range of substrates. Other phosphines have to be tried, with maybe less steric hindrance, or free carbenes could also be used to activate the triple bond of the alkyne. Finally, a catalyst

Summary and future aspects

based on a metal such as gold or platinum could be the solution of this challenge since the Markovnikov addition of other hydrogen bonded nucleophiles (i.e. water, amines, carboxylic acids) was often found to be catalyzed by complexes containing these metals.

4 Acknowledgements

This work was carried out in the Department of Organic Chemistry of the Technische Universität Kaiserslautern, Germany.

I would like to express my sincere gratitude and thanks to all colleagues, past and present, in the department for making the time I spent at the TU KL very enjoyable and unforgettable.

I particularly would like to express my special gratitude to Prof. Dr. Lukas J. Gooßen, for being a superb supervisor and for his inspiring guidance, never failing enthusiasm throughout this study, for sharing his wisdom in science and life and for providing excellent working facilities. Also for introducing me to the world of catalysis and ruthenium chemistry.

I would like to thank all members of the AK Gooßen:

My co-authors: Claus Brinkmann, Kifah M. Salih, Matthias Arndt for fruitful collaboration in the enamide project. Paul Lange, Heiko Bauer, Marcus Fischer, Patrizia Mamone for their fine contributions to the projects during their training stages.

My co-workers and room mates: Thomas Knauber (Der Knauber), Dr. Corneliu Stanciu (thank you for the brilliant linguistic revision of my thesis) and Roman Bauer (flash-games Meister) for valuable discussions and very pleasant times in the lab.

My proofreaders: Dr Nuria Rodríguez Garrido, Dominik Ohlmann (thank you for your help on document formatting), Bettina Zimmermann née Melzer (Schatzelein) for constructive criticism and linguistic revision of the thesis.

Felix Rudolphi and Andreas Fromm, for excellent assistance in computer matters.

Filipe Costa, Bilal Ahmad Khan, Christoph Oppel and finally Christophe Linder (my colleague, flat mate and friend) for a good time both within and outside the lab.

I also would like to thank all the personnel of the University of Kaiserslautern, Dr. Harald Kelm and Christiane Müller for their expertise with the NMR-machines and in theoretical

Acknowledgements

NMR problems. Ludvik Napast, Jürgen Rahm, Frank Schröer, Csaba Tömör and Carmen Eggert for always being there when I needed help.

I would like to thank my family (thank you for forgiving me for moving to Germany),my mother Murielle Blanchot, my sister Emilie Blanchot, my friends (for always being supportive) and my girlfriend Irina who had to support me during all this long and stressful writing period.

Finally I also would like to thank the Gesellschaft Deutscher Chemiker (GDCh) for making it possible for me to attend several meetings and courses. Financial support was obtained from Deutsche Forschungsgemeinschaft (DFG) and Umicore AG.

5 Experimental part

5.1 General information

5.1.1 Solvents and chemicals

All used solvents were purified by general described procedures.[94] All commercially available reagents were distilled either from CaH$_2$ or molecular sieves under an inert atmosphere before use. All commercially available imides were used as received. All commercially available thioamides were used a received, while all others were obtained by converting the corresponding amides into thioamides using the method of Lawesson et al.[88c] All commercial primary amides were used as obtained without further purification, while the other were synthesized by treating the commercially corresponding carboxylic acid with oxalyl chloride and ammonia solution in dry methylene chloride. All commercial amides were used as obtained without further purification. [Ru(cod)(met)$_2$] was supplied by Umicore AG.

5.1.2 Analytical methods

Nuclear Magnetic Resonance

Proton- and decoupled carbon-NMR spectra were recorded with a Bruker FT-NMR DPX 200, DPX 400 and a Bruker Avance 600. The frequency and solvent used is described separately for each substance. Chemical shifts are given in units of the δ-scale in ppm. Shifts for ^{1}H-spectra are given respectively to the proton signal of the solvent used (chloroform: 7.25 ppm, dimethyl sulfoxide: 2.50 ppm, methanol: 3.35 ppm, water: 4.75 ppm), for ^{13}C-spectra respectively to the deuterated solvent (chloroform: 77.0 ppm, dimethyl sulfoxide: 37.7 ppm, methanol: 49.3 ppm). *The atom numbering within products is not according to the IUPAC rules.* The multiplicity of the signals is abbreviated by the following letters:

Coupling constants are given in Hertz (Hz). Processing and interpretation was performed with ACD-Labs 7.0 (Advanced Chemistry Development Inc.)

abbreviation	signal
s	singlet
d	doublet
dd	doublet of doublet
ddd	doublet of doublet of doublet
td	triplet of doublet
q	quartet

Gaschromatography

For GC-analysis a Hewlett Packard 6980 chromatograph was used. The gas carrier was nitrogen with a flow rate of 149mL/min (0.5 bar pressure). The temperature of the injector was 220 °C. The split-ratio was 1:100. For separation an Agilent HP-5-column with 5% phenyl-methyl-siloxane (30 x 320 µm x 1.0 µm, 100/ 2.3-30-300/ 3) was used. The following temperature program was implemented: starting temperature 60 °C (2min), linear temperature increase (30 °C min^{-1}) to 300 °C, end temperature 300 °C (13min).

High-Performance-Liquid-Chromatography

A Shimadzu High Performance Liquid Chromatograph with a Merck Reversed Phase LiChroCat© PAH C 18 column with a particle diameter of 5 µm was used at a constant oven temperature of 60°C and a column pressure of 125 bar. The eluents used were acetonitrile and water with a flow rate of 1mL/min. Gradient: 15% acetonitrile for 2min, linear increase to 85% acetonitrile during the course of 8min, constant gradient for 3.5min, decrease to 15% within 0.1min and constant for 3min. The standard configuration injected 10 µL of the sample. This amount could be manually changed with the control and interpretation program Shimadzu Class-VP.

Mass Spectrometry

Mass spectrometry was performed with a GC-MS Varian Saturn 2100 T. The ionization was done by EI AGC. The intensities of the signals are relative to the highest peak. For fragments with isotopomers only the more intensive peak of the isotopomer is given.

Experimental part

Infrared Spectroscopy

Infrared spectra were recorded with a Perkin-Elmer Fourier Transform Infrared Spectrometer FT/IR. Solids were thoroughly ground and mixed with potassium bromide and pressed into a pellet. Liquids were measured as a thin film in between sodium chloride plates. Absorbance-bands are shown in wave numbers (cm^{-1}). Intensities are abbreviated: s (strong), m (medium) and b (broad).

Elemental Analysis

CHN-elemental analysis was performed with a Perkin-Elmer Elemental Analyzer EA 2400 CHN.

5.1.3 High-throughput experiments

In order to perform a vast number of experiments a specially manufactured setup was used. All reactions were carried out in 20 mL headspace vials that were closed and clamped shut with aluminum caps fitted with a Teflon-coated butyl rubber septum (both commercially available at Macherey & Nagel).

In 8 cm high round aluminum cases, which fit the hot plate of a regular laboratory heater in diameter, 10 of the thus equipped 20 mL headspace vials can be tempered between 25°C and 180 °C. An 11th smaller hole drilled in the middle of the case creates room to hold the thermometer of the heater (Figure 4 left).

To correctly evacuate and refill 10 reaction vessels with inert gas at the same time, special vacuum distributors were manufactured to be connected to the Schlenck-line (Figure 4 right). A steel tubing is linked to ten 3 mm Teflon tubes, which are equipped on the opposite end with adaptors for Luer-Lock syringe needles. The steel tubing can be connected to the Schlenck-line just like any other laboratory equipment by a steel olive and vacuum tubing.

To perform 10 or more reactions parallel the following protocol was used. All solid substances were weighed in the reaction vessels, an oven-dried, hot 20 mm stir bar added and each vessel closed with a separate cap using flanging pliers. All 10 vessels were transferred to one of the aforementioned aluminum cases and evacuated using syringe needles connected to the vacuum steel tubing.

Experimental part

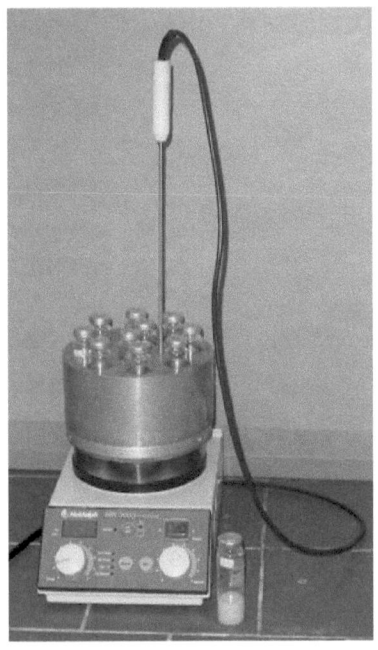 

Figure 4: 10-vessel aluminium case with heater (left) reaction setup and vacuum distributor (right).

The reaction vessels were evacuated and refilled with nitrogen. An oil bubbling valve at the top of the steel tubing was used to guarantee a pressure release. Using standard sterile and Hamilton syringes all liquid reagents, stock solutions of reagents, solvents and the standard (n-tetradecane) were added and the vessels were evacuated and refilled with nitrogen 3 times. After removal of the needles, the aluminum case was tempered to the desired temperature. Every temperature description is the case temperature, which only differs by maximum 2°C from the actual reaction media temperature.

At the end of the reaction time and after cooling to room temperature, the reaction vials were opened carefully. 2 mL of ethyl acetate were added to dilute the reaction mixture and with a disposable pipette mixed thoroughly to ensure a homogenous mixture. A 0.25 mL sample was withdrawn and extracted with 2 mL of ethyl acetate and 2 mL of brine-solution or 2 mL of saturated potassium carbonate solution. The organic layer was filtered through a pipette filled with a cotton plug and magnesium sulfate into a GC-vial.

Experimental part

After evaluating the contents on the GC and if necessary GC-MS, the contents of all work-up and analysis vials were recombined and the product isolated using standard procedures, deposed on silica-gel and purified by flash chromatography using a *Combi Flash Companion-Chromatography* apparatus from Ico-Systems (Figure 5).

Figure 5: *Combi Flash Companion-Chromatography.*

The developed experimental setups and an electronic laboratory journal allowed a substantial amount of reactions to be performed during the course of this work. Approximately 2500 reactions would have consumed a much longer time using standard laboratory techniques. Preparative reactions were performed mostly in standard laboratory oven-dried glass ware. The following experimental section describes all reactions performed, that are mentioned in the theoretical section above. Yields are isolated yields if nothing else is mentioned. All known compounds were analyzed by proton and carbon NMR, as well as GC-MS when possible and elemental analysis.

5.2 Addition of imides to alkynes

5.2.1 General method

Method A: An oven-dried flask was charged with the N-nucleophile (1.00 mmol), bis(2-methallyl)-cycloocta-1,5-diene-ruthenium(II) (6.4 mg, 0.02 mmol) and scandium triflate (19.7 mg, 0.04 mmol) and flushed with nitrogen. Subsequently, dry DMF (3.0 mL), tri-n-butyl phosphine (15 µL, 0.06 mmol) and alkyne (2.00 mmol) were added *via* syringe. The resulting solution was stirred for 15 h at 60 °C, and then poured into an aqueous sodium bicarbonate solution (30 mL). The resulting mixture was extracted repeatedly with 20 mL portions of ethyl acetate, the combined organic layers were washed with water and brine, dried with magnesium sulfate, filtered, and the volatiles were removed in vacuo. The residue was purified by column chromatography (silica gel, ethyl acetate/hexane) to yield the Z-enimide. The identity and purity of the products were confirmed by ^{1}H and ^{13}C NMR spectroscopy, mass spectroscopy and elemental analysis.

Method B: An oven-dried flask was charged with the N-nucleophile (1.00 mmol), bis(2-methallyl)-cycloocta-1,5-diene-ruthenium(II) (16 mg, 0.05 mmol) and scandium triflate (19.7 mg, 0.04 mmol) and flushed with nitrogen. Subsequently, dry DMF (3.0 mL), tri-isopropylphosphine (29 µL, 0.15 mmol) and alkyne (2.00 mmol) were added *via* syringe. The resulting solution was stirred for 15 h at 60 °C, and then poured into an aqueous sodium bicarbonate solution (30 mL). The resulting mixture was extracted repeatedly with 20 mL portions of ethyl acetate, the combined organic layers were washed with water and brine, dried with magnesium sulfate, filtered, and the volatiles were removed in vacuo. The residue was purified by column chromatography (silica gel, ethyl acetate/hexane) to yield the E-enimide. The identity and purity of the products were confirmed by ^{1}H and ^{13}C NMR spectroscopy, mass spectroscopy and elemental analysis.

Experimental part

5.2.2 Synthesis of the enimides

Synthesis of N-[(Z)-hex-1-enyl]succinimide (3aa)

3aa was synthesized following method A using succinimide (99.1 mg, 1.00 mmol) and 1-hexyne (229 μL, 2.0 mmol) and purified by column chromatography (1:3 ethyl acetate/hexane), yielding the product as a yellowish oil (177.6 mg, 98%).

^{1}H-NMR   (400 MHz, CDCl$_3$, 25 °C): δ=5.82 (d, ^3J = 8.6 Hz, 1H, H-5), 5.68 (dt, ^3J = 15.0 Hz, 7.5 Hz, 1H, H-6), 2.74 (s, 4H, H-1, H-2), 1.89 (dt, J = 14.6, 7.3, 1.6 Hz, 2H, H-7), 1.25 – 1.36 (m, 4H, H-8, H-9), 0.83 (t, J = 7.1 Hz, 3H, H-10) ppm.

^{13}C-NMR   (101 MHz, CDCl$_3$, 25 °C): δ=175.6 (C-3, C-4), 133.6 (C-5), 116.7 (C-6), 30.7 C-7), 28.4 (C-1, C-2), 27.7 (C-8), 22.3 (C-9), 13.8 (C-10) ppm.

EA   Calcd: C, 66.2; H, 8.3; N, 7.7.
Found: C, 66.3; H, 8.4; N, 7.7.

MS   (Ion trap, EI): m/z (%) = 182 (13, [M$^+$]), 138 (67), 100 (75), 82 (100), 56 (44).

Synthesis of N-[(Z)-hex-1-en-1-yl]glutarimide (3ba)

3ba was synthesized following method A using glutarimide (113.1 mg, 1.00 mmol) and 1-hexyne (229 μL, 2.0 mmol) and purified by column chromatography (1:3 ethyl acetate/hexane), yielding the product as a colorless oil (113.3 mg, 58%).

^{1}H-NMR   (400 MHz, CDCl$_3$, 25 °C): δ=5.88 (d, ^3J(H,H)=8.3 Hz, 1H, H-6), 5.70 (dt, ^3J(H,H)=15.0 Hz, 7.5 Hz, 1H, H-7), 2.67 (t, ^3J(H,H)=6.66 Hz, 4H, H-2, H-5), 1.96 (t, ^3J(H,H)=7.3 Hz, 2H, H-1), 1.78 (dt, ^3J(H,H)=14.5, 7.3, 1.56 Hz, 2H, H-8) 1.23 – 1.33 (m, 4H, H-9, H-10), 0.82 (t, ^3J(H,H)=7.14 Hz, 3H, H-11) ppm.

^{13}C-NMR   (101 MHz, CDCl$_3$, 25 °C): δ=171.6 (C-3, C-4), 132.9 (C-6), 120.0 (C-7), 32.9 (C-2, C-5), 30.3 (C-8), 26.9 (C-9), 22.2 (C-10), 17.2 (C-1), 13.7 (C-11) ppm.

EA   Calcd: C, 67.7; H, 8.8; N, 7.2.
Found: C, 66.9; H, 8.7; N, 7.0.

MS   (Ion trap, EI): m/z (%) = 195 (100, [M$^+$]), 138 (52), 114 (69), 86 (40), 56 (50).

Experimental part

Synthesis of N-[(Z)-hex-1-en-1-yl]-3,3-dimethylglutarimide (3ca)

3ca was synthesized following method A using 3,3-dimethylglutarimide (141.2 mg, 1.00 mmol) and 1-hexyne (229 µL, 2.0 mmol) and purified by column chromatography (1:3 ethyl acetate/hexane), yielding the product as a yellowish oil (138.5 mg, 62%).

¹H-NMR (400 MHz, CDCl₃, 25 °C): δ=5.90 (d, ³J(H,H)=8.22 Hz, 1H, H-6), 5.74 (dt, ³J(H,H)=15.0 Hz, 8.22 Hz, H-7), 2.56 (s, 4H, H-2, H-5), 1.83 (dt, ³J(H,H)=14.5, 7.29, 1.61 Hz, 2H, H-8), 1.23 − 1.39 (m, 4H, H-9, H-10), 1.13 (s, 6H, H-12, H-13), 0.85 (t, ³J(H,H)=7.1 Hz, 3H, H-11) ppm.

¹³C-NMR (101 MHz, CDCl₃, 25 °C): δ=171.0 (C-3, C-4), 133.2 (C-6), 119.7 (C-7), 46.5 (C-12, C-13), 30.4 (C-8), 29.2 (C-1), 27.8 (C-9), 27.2 (C-2, C-5), 22.3 (C-10), 13.7 (C-11) ppm.

EA Calcd: C, 69.9; H, 9.5; N, 6.3.
Found: C, 69.8; H, 9.6; N, 6.3.

MS (Ion trap, EI): m/z (%) = 224 (100, [M⁺]), 180 (32), 142 (35), 83 (42), 56 (21).

Synthesis of N-[(Z)-hex-1-en-1-yl]phthalimide (3da)

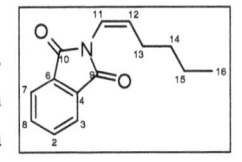

3da was synthesized following method A using phtalimide (147.2 mg, 1.00 mmol) and 1-hexyne (229 µL, 2.0 mmol) and purified by column chromatography (1:3 ethyl acetate/hexane), yielding the product as a white solid (201.8 mg, 88 %).

¹H-NMR (400 MHz, CDCl₃, 25 °C): δ=7.79 − 7.88 (m, 2H, H_arom), 7.67 − 7.74 (m, 2H, H_arom), 6.05 (dt, ³J(H,H)=8.41, 1.66 Hz, 1H, H-11), 5.70 − 5.77 (m, 1H, H-12), 2.04 (dt, ³J(H,H)=14.55, 7.3, 1.6 Hz, 2H, H-13), 1.25 − 1.43 (m, 4H, H-14, H-15), 0.83 (t, ³J(H,H)=7.24 Hz, 3H, H-16) ppm.

¹³C-NMR (101 MHz, CDCl₃, 25 °C): δ=166.7 (C-9, C-10), 134.2 (C_arom), 133.5 (C_arom), 132.1 (C-11), 123.4 (C_arom), 116.2 (C-12), 30.8 (C-13), 28.0 (C-14), 22.2 (C-15), 13.9 (C-16) ppm.

EA Calcd: C, 73.3; H, 6.6; N, 6.1.
Found: C, 73.2; H, 6.5; N, 6.1.

MS (Ion trap, EI): m/z (%) = 229 (8, [M⁺]), 186 (100), 148 (20), 76 (14), 56 (15).

Synthesis of N-[(E)-hex-1-enyl]-3,4-pyridindicarboximid (3ea)

3ea was synthesized following method A using 3,4-pyridinedicarboximide (148.1 mg, 1.00 mmol) and 1-hexyne (229 µL, 2.0 mmol) and purified by column chromatography (1:3 ethyl acetate/hexane), yielding the product as a yellowish oil (13.5 mg, 14 %).

^{1}H-NMR (400 MHz, CDCl$_3$, 25 °C): δ=9.18 (s, 1H, H$_{arom}$), 9.08 (d, ^3J(H,H)=4.8 Hz, 1H, H$_{arom}$), 7.77 (d, ^3J(H,H)=5.2 Hz, 1H, H$_{arom}$), 6.09 (d, ^3J(H,H)=8.2 Hz, 1H, H-9), 5.85 (m, 1H, H-10), 2.04 (dt, ^3J(H,H)=14.8, 7.4, 1.61 Hz, 2H, H-11), 1.25 – 1.47 (m, 4H, H-12, H-13), 0.86 (t, ^3J(H,H)=7.2 Hz, 3H, H-14) ppm.

^{13}C-NMR (101 MHz, CDCl$_3$, 25 °C): δ=165.7 (C-7), 165.3 (C-8), 155.8 (C$_{arom}$), 145.1 (C$_{arom}$), 139.5 (C$_{arom}$), 134.4 (C-9), 124.8 (C$_{arom}$), 116.9 (C-10), 115.7 (C$_{arom}$), 30.8 (C-11), 28.0 (C-12), 22.3 (C-13), 13.8 (C-14) ppm.

EA Calcd: C, 67.81; H, 6.13; N, 12.17.
Found: C, 68.15; H, 6.17; N, 12.02.

MS (Ion trap, EI): m/z (%) = 230 (8, [M$^+$]), 187 (100), 150 (24), 82 (18), 50 (22).

Synthesis of N-acetyl-N-hex-1-enyl-acetamide (3fa)

3fa was synthesized following method A using diacetamide (101.1 mg, 1.00 mmol) and 1-hexyne (229 µL, 2.0 mmol) and yield was determined by GC using n-tetradecane as internal standard (8%).

MS (Ion trap, EI): m/z (%) = 182 (93, [M$^+$]), 139 (100), 125 (34), 83 (42), 43 (52).

Experimental part

Synthesis of N-[(Z)-hex-1-enyl]-1,5,5-trimethylhydantoin (3ga)

3ga was synthesized following method A using 1,5,5-trimethylhydantoin (156.2 mg, 1.00 mmol) and 1-hexyne (229 μL, 2.0 mmol) and purified by column chromatography (1:3 ethyl acetate/hexane), yielding the product as a yellowish oil (201.9 mg, 90 %).

¹H-NMR	(400 MHz, CDCl$_3$, 25 °C): δ=5.79 (d, ^3J(H,H)=8.41 Hz, 1H, H-7), 5.54 (dt, ^3J(H,H)=15.1 Hz, 7.56 Hz, 1H, H-8), 2.80 (s, 3H, H-1), 1.87 (dt, ^3J(H,H)=14.6, 7.3, 1.62 Hz, 2H, H-9), 1.34 (s, 6H, H-2, H-3), 1.13 – 1.33 (m, 4H, H-10, H-11), 0.75 (t, ^3J(H,H)=7.24 Hz, 3H, H-12) ppm.
¹³C-NMR	(101 MHz, CDCl$_3$, 25 °C): δ=174.7 (C-6), 154.3 (C-5), 131.7 (C-7), 116.8 (C-8), 61.5 (C-4), 30.7 (C-9), 27.9 (C-10), 24.5 (C-1), 22.5 (C-11), 22.1 (C-2, C-3), 13.7 (C-12) ppm.
EA	Calcd: C, 64.2; H, 8.9; N, 12.5. Found: C, 64.0; H, 8.7; N, 12.3.
MS	(Ion trap, EI): m/z (%) = 224 (100, [M⁺]), 194 (18), 142 (36), 83 (42), 56 (22).

Synthesis of 3-hex-1-enyl-1-methyl-imidazolidine-2,4-dione (3ha)

3ha was synthesized following method A using 1-methylhydantoine (114.1 mg, 1.00 mmol) and 1-hexyne (229 μL, 2.0 mmol) and yield was determined by GC using n-tetradecane as internal standard (8%).

MS (Ion trap, EI): m/z (%) = 197 (18, [M⁺]), 153 (45), 124 (53), 82 (54), 42 (22).

Synthesis of 3-hex-1-enyl-thiazolidine-2,4-dione (3ia)

3ia was synthesized following method A using thiazolidine-2,4-dione (117.1 mg, 1.00 mmol) and 1-hexyne (229 μL, 2.0 mmol) and yield was determined by GC using n-tetradecane as internal standard (16%).

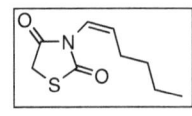

MS (Ion trap, EI): m/z (%) = 200 (4, [M⁺]), 156 (16), 128 (28), 118 (21), 101 (23).

Experimental part

Synthesis of N-((E)-hex-1-enyl)-1-methyluracile (3ja)

3ja was synthesized following method A using 1-methyluracile (126.1 mg, 1.00 mmol) and 1-hexyne (229 µL, 2.0 mmol) and yield was determined by GC using n-tetradecane as internal standard (4%).

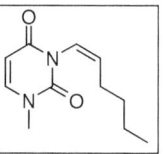

MS (Ion trap, EI): m/z (%) = 208 (12, [M+]), 179 (53), 166 (29), 151 (57), 127 (100).

Synthesis of N-[(Z)-2-phenylvinyl]succinimide (3ab)

3ab was synthesized following method A using succinimide (99.1 mg, 1.00 mmol) and phenylacetylene (220 µL, 2.0 mmol) and purified by column chromatography (1:3 ethyl acetate/hexane), yielding the product as a white solid (114.7 mg, 57 %).

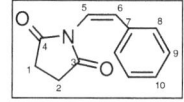

¹H-NMR (400 MHz, CDCl$_3$, 25 °C): δ=7.19 – 7.29 (m, 3H, H$_{arom}$), 7.15 (d, ³J(H,H)=7.1 Hz, 2H, H$_{arom}$), 6.65 (d, ³J(H,H)=9.3 Hz, 1H, H-5), 6.07 (d, ³J(H,H)=9.3 Hz, 1H, H-6), 2.68 (s, 4H, H-1, H-2) ppm.

¹³C-NMR (101 MHz, CDCl$_3$, 25 °C): δ=175.2 (C-3, C-4), 134.7 (C-5), 131.0 (C$_{arom}$), 128.5 (C$_{arom}$), 128.2 (C$_{arom}$), 128.0 (C$_{arom}$), 117.1 (C-6), 28.5 (C-1, C-2) ppm.

EA Calcd: C, 73.3; H, 6.6; N, 6.1.
Found: C, 73.3; H, 6.6; N, 6.0.

MS (Ion trap, EI): m/z (%) = 229 (5, [M+]), 138 (100), 110 (17), 65 (21), 56 (32).

Synthesis of N-[(Z)-4-phenylbut-1-en-1-yl]succinimide (3ac) (CAS-N0: 919082-94-3)

3ac was synthesized following method A using succinimide (99.1 mg, 1.00 mmol) and 4-phenyl-1-butyne (281 µL, 2.0 mmol) and purified by column chromatography (1:3 ethyl acetate/hexane), yielding the product as a white solid (226.3 mg, 99 %).

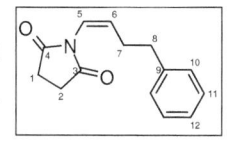

¹H-NMR (400 MHz, CDCl$_3$, 25 °C): δ=7.21 – 7.24 (m, 2H, H$_{arom}$), 7.12 – 7.16 (m, 3H, H$_{arom}$), 5.87 (d, ³J(H,H)=8.31 Hz, 1H, H-5), 5.70 (d, ³J(H,H)=8.41 Hz, 1H, H-

Experimental part

6), 2.70 (t, ^3J(H,H)=7.8 Hz, 2H, H-8), 2.67 (s, 4H, H-1, H-2), 2.24 (dt, ^3J(H,H)=14.6, 7.3, 1.6 Hz, 2H, H-7) ppm.

^{13}C-NMR (101 MHz, CDCl$_3$, 25 °C): δ=175.5 (C-4, C-5), 141.3 (C$_{arom}$), 131.8 (C-5), 128.5 (C$_{arom}$), 126.0 (C$_{arom}$), 117.4 (C-6), 34.7 (C-7), 30.0 (C-8), 28.4 (C-1, C-2) ppm.

EA Calcd: C, 71.6; H, 5.5; N, 7.0.
Found: C, 71.5; H, 5.6; N, 6.7.

MS (Ion trap, EI): *m/z* (%) = 201 (100, [M$^+$]), 172 (5), 118 (65), 89 (25), 55 (15).

The data correspond to those reported in the literature: T. Ogawa, T. Kiji, K. Hayami, H. Suzuki, *Chem. Lett.* **1991**, *20*, 1443.

Synthesis of *N*-[(Z)-oct-1-en-1-yl]succinimide (3ad)

3ad was synthesized following method A using succinimide (99.1 mg, 1.00 mmol) and 1-octyne (298 µL, 2.0 mmol) and purified by column chromatography (1:3 ethyl acetate/hexane), yielding the product as a yellowish oil (73.4 mg, 74 %).

^{1}H-NMR (400 MHz, CDCl$_3$, 25 °C): δ=5.81 (d, ^3J(H,H)=8.51 Hz, 1H, H-5), 5.67 (dt, ^3J(H,H)=15.1 Hz, 7.5 Hz, 1H, H-6), 2.73 (s, 4H, H-1, H-2), 1.87 (qd, ^3J(H,H)=7.38, 1.61 Hz, 2H, H-7), 1.29 – 1.38 (m, 4H, H-7, H-8), 1.16 – 1.27 (m, 6H, H-9, H-10, H-11), 0.81 (t, ^3J(H,H)=7.1 Hz, 3H, H-12) ppm.

^{13}C-NMR (101 MHz, CDCl$_3$, 25 °C): δ=175.6 (C-3, C-4), 133.7 (C-5), 116.7 (C-6), 31.6 (C-7), 28.8 (C-8), 28.4 (C-1, C-2), 28.1 (C-9), 22.54 (C-10), 22.51 (C-11), 14.0 (C-12) ppm.

EA Calcd: C, 68.9; H, 9.15; N, 6.7.
Found: C, 68.8; H, 9.17; N, 6.6.

MS (Ion trap, EI): *m/z* (%) = 210 (10, [M$^+$]), 138 (70), 110 (100), 81 (48), 55 (40).

Synthesis of *N*-[(Z)-5-chloropent-1-en-1-yl]succinimide (3ae)

3ae was synthesized following method A using succinimide (99.1 mg, 1.00 mmol) and 5-chloro-1-pentyne (213 µL, 2.0 mmol) and purified

- 138 -

by column chromatography (1:3 ethyl acetate/hexane), yielding the product as a yellowish oil (115.5 mg, 57%).

^{1}H-NMR (400 MHz, CDCl$_3$, 25 °C): δ=5.91 (d, ^{3}J(H,H)=8.61 Hz, 1H, H-5), 5.65 (m, 1H, H-6), 3.47 (t, ^{3}J(H,H)=7.1, 2H, H-9), 2.75 (s, 4H, H-1, H-2), 2.09 (m, 2H, H-7), 1.83 (m, 2H, H-8) ppm.

^{13}C-NMR (101 MHz, CDCl$_3$, 25 °C): δ=175.5 (C-3, C-4), 131.2 (C-5), 118.4 (C-6), 44.2 (C-9), 31.1 (C-7), 28.4 (C-1, C-2), 25.1 (C-8) ppm.

EA Calcd: C, 53.6; H, 6.0; N, 6.9.
Found: C, 53.8; H, 6.1; N, 7.1.

MS (Ion trap, EI): m/z (%) = 202 (20, [M$^+$]), 166 (48), 138 (100), 82 (45), 56 (38).

Synthesis of 1-(3-oxo-but-1-enyl)-pyrrolidine-2,5-dione (3af)

3af was synthesized following method A using succinimide (99.1 mg, 1.00 mmol) and but-3-yn-2-one (179 µL, 2.0 mmol) and yield was determined by GC using n-tetradecane as internal standard (16%).

MS (Ion trap, EI): m/z (%) = 183 (10, [M$^+$]), 152 (52), 124 (100), 96 (24), 70 (19), 52 (23).

Synthesis of 1-(3-methyl-buta-1,3-dienyl)-pyrrolidine-2,5-dione (3ag)

3ag was synthesized following method A using succinimide (99.1 mg, 1.00 mmol) and 2-methyl-but-1-en-3-yne (192 µL, 2.0 mmol) and yield was determined by GC using n-tetradecane as internal standard (16%).

MS (Ion trap, EI): m/z (%) = 165 (64, [M$^+$]), 122 (52), 81 (100), 69 (22), 55 (44), 41 (20).

Experimental part

Synthesis of 1-(3,3-dimethyl-but-1-enyl)-pyrrolidine-2,5-dione (3ah)

3ah was synthesized following method A using succinimide (99.1 mg, 1.00 mmol) and 3,3-dimethyl-but-1-yne (192 µL, 2.0 mmol) and yield was determined by GC using *n*-tetradecane as internal standard (8%).

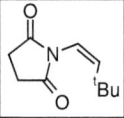

MS   (Ion trap, EI): *m/z* (%) = 181 (37, [M$^+$]), 166 (100), 148 (17), 138 (13), 120 (19).

Synthesis of 1-(3-methoxy-propenyl)-pyrrolidine-2,5-dione (3ai)

3ai was synthesized following method A using succinimide (99.1 mg, 1.00 mmol) and 3-methoxypropyne (169 µL, 2.0 mmol) and yield was determined by GC using *n*-tetradecane as internal standard (6%).

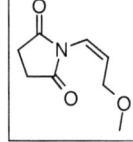

MS   (Ion trap, EI): *m/z* (%) = 170 (82, [M$^+$]), 158 (17), 141 (100), 77 (67), 51 (22).

Synthesis of *N*-[(Z)-2-phenylvinyl]phthalimide (3db) (CAS-N0: 136613-13-3)

3db was synthesized following method A using phtalimide (147.2 mg, 1.00 mmol) and phenylacetylene (220 µL, 2.0 mmol) and purified by column chromatography (1:3 ethyl acetate/hexane), yielding the product as a white solid (132,4 mg, 53 %).

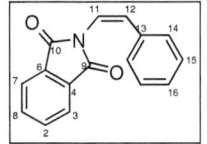

^{1}H-NMR   (400 MHz, CDCl$_3$, 25 °C): δ=7.86 (dd, ^3J(H,H)=5.48, 3.13 Hz, 2H, H$_{arom}$), 7.74 (dd, ^3J(H,H)=5.48, 3.13 Hz, 2H, H$_{arom}$), 7.19 – 7.28 (m, 5H, H$_{arom}$), 6.70 (d, ^3J(H,H)=9.19 Hz, 1H, H-11), 6.33 (d, ^3J(H,H)=9.19 Hz, 1H, H-12) ppm.

^{13}C-NMR   (101 MHz, CDCl$_3$, 25 °C): δ=166.3 (C$_{arom}$), 135.1 (C$_{arom}$), 134.3 (C-11), 132.2 (C$_{arom}$), 130.2 (C$_{arom}$), 128.4 (C$_{arom}$), 128.2 (C$_{arom}$), 128.0 (C$_{arom}$), 123.7 (C$_{arom}$), 116.4 (C-12) ppm.

EA   Calcd: C, 77.1; H, 4.4; N, 5.6.
Found: C, 76.5; H, 4.45; N, 5.2.

MS   (Ion trap, EI): *m/z* (%) = 249 (100, [M$^+$]), 204 (22), 104 (25), 76 (42), 50 (35).

The data correspond to those reported in the literature: T. Ogawa, T. Kiji, K. Hayami, H. Suzuki *Chem. Lett.* **1991**, *20*, 1443.

Experimental part

Synthesis of N-[(Z)-2-phenylvinyl]-1,5,5-trimethylhydantoine (3gb)

3gb was synthesized following method A using 1,5,5-trimethylhydantoin (156.2 mg, 1.00 mmol) and phenylacetylene (220 µL, 2.0 mmol) and purified by column chromatography (1:3 ethyl acetate/hexane), yielding the product as a yellowish oil (80.6 mg, 33 %).

¹H-NMR	(400 MHz, CDCl₃, 25 °C): δ=7.30 (d, ³*J*(H,H)=7.6 Hz, 2H, H$_{arom}$), 7.26 (t, ³*J*(H,H)=7.4 Hz, 1H, H$_{arom}$), 7.22 (d, ³*J*(H,H)=7.4 Hz, 2H, H$_{arom}$), 6.65 (d, ³*J*(H,H)=8.8 Hz, 1H, H-7), 6.19 (d, ³*J*(H,H)=9.0 Hz, 1H, H-8), 2.92 (s, 3H, H-1), 1.36 (s, 6H, H-2, H-3) ppm.
¹³C-NMR	(101 MHz, CDCl₃, 25 °C): δ=173.7 (C-5), 154.0 (C-6), 135.0 (C-7), 130.1 (C$_{arom}$), 128.3 (C$_{arom}$), 128.1 (C$_{arom}$), 128.0 (C$_{arom}$), 116.9 (C-8), 61.4 (C-4), 24.7 (C-1), 22.0 (C-2, C-3) ppm.
EA	Calcd: C, 74.9; H, 7.1; N, 11.2. Found: C, 74.5; H, 6.7; N, 11.2.
MS	(Ion trap, EI): *m/z* (%) = 244 (100, [M⁺]), 145 (62), 117 (15), 89 (14), 56 (41).

Synthesis of N-[(Z)-2-phenylvinyl]-4,4-dimethylglutarimide (3cb)

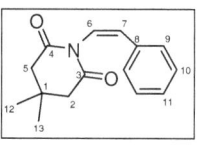

3cb was synthesized following method A using 3,3-dimethylglutarimide (141.2 mg, 1.00 mmol) and phenylacetylene (220 µL, 2.0 mmol) and purified by column chromatography (1:3 ethyl acetate/hexane), yielding the product as a yellowish oil (121.7 mg, 50 %).

¹H-NMR	(400 MHz, CDCl₃, 25 °C): δ=7.28 (d, ³*J*(H,H)=7.8 Hz, 2H, H$_{arom}$), 7.23 (t, ³*J*(H,H)=7.4 Hz, 1H, H$_{arom}$), 7.18 (d, ³*J*(H,H)=7.1 Hz, 2H, H$_{arom}$), 6.76 (d, ³*J*(H,H)=8.8 Hz, 1H, H-6), 6.16 (d, ³*J*(H,H)= 9.0 Hz, 1H, H-7), 2.50 (s, 4H, H-2, H-5), 1.06 (s, 6H, H-12, H-13) ppm.
¹³C-NMR	(101 MHz, CDCl₃, 25 °C): δ=171.3 (C-3, C-4), 134.8 (C-6), 131.8 (C$_{arom}$), 128.4 (C$_{arom}$), 128.0 (C$_{arom}$), 127.8 (C$_{arom}$), 120.9 (C-7), 46.5 (C-2, C-5), 29.4 (C-1), 27.9 (C-12, C-13) ppm.

EA	Calcd: C, 74.0; H, 7.0; N, 5.8.
	Found: C, 73.9; H, 6.8; N, 5.7.
MS	(Ion trap, EI): m/z (%) = 243(100, [M⁺]), 200 (80), 119 (35), 83 (38), 55 (22).

Synthesis of N-[(Z)-2-phenylvinyl]glutarimide (3bb)

3bb was synthesized following method A using glutarimide (113.1 mg, 1.00 mmol) and phenylacetylene (220 µL, 2.0 mmol) and purified by column chromatography (1:3 ethyl acetate/hexane), yielding the product as a white solid (129.2 mg, 60%).

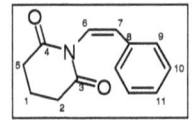

¹H-NMR (400 MHz, CDCl₃, 25 °C): δ=7.28 (t, ³J(H,H)=7.8 Hz, 2H, H_arom), 7.24 (t, ³J(H,H)=7.4 Hz, 1H, H_arom), 7.18 (d, ³J(H,H)=7.4 Hz, 2H, H_arom), 6.71 (d, ³J(H,H)=9.0 Hz, 1H, H-6), 6.18 (d, ³J(H,H)=8.8 Hz, 1H, H-7), 2.64 (t, ³J(H,H)=6.5 Hz, 4H, H-2, H-5), 1.92 – 1.97 (m, 2H, H-1) ppm.

¹³C-NMR (101 MHz, CDCl₃, 25 °C): δ=171.8 (C-3, C-4), 134.7 (C-6), 131.2 (C_arom), 128.5 (C_arom), 128.1 (C_arom), 127.8 (C_arom), 121.0 (C-7), 33.0 (C-2, C-5), 17.0 (C-1) ppm.

EA	Calcd: C, 72.5; H, 6.0; N, 6.3.
	Found: C, 72.3; H, 5.7; N, 6.3.
MS	(Ion trap, EI): m/z (%) = 196 (100, [M⁺]), 166 (22), 138 (52), 86 (40), 56 (50).

Synthesis of N-[(Z)-4-phenylbut-1-en-1-yl]-1,5,5-trimethylhydantoine (3gc)

3gc was synthesized following method A using 1,5,5-trimethylhydantoin (156.2 mg, 1.00 mmol) and 4-phenyl-1-butyne (281 µL, 2.0 mmol) and purified by column chromatography (1:3 ethyl acetate/hexane), yielding the product as a yellowish oil (256.1 mg, 94 %).

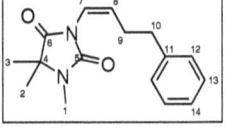

¹H-NMR (400 MHz, CDCl₃, 25 °C): δ=7.27 (t, ³J(H,H)=7.4 Hz, 2H, H_arom), 7.19 (t, ³J(H,H)=7.8 Hz, 3H, H_arom), 5.97 (d, ³J(H,H)=8.6 Hz, 1H, H-7), 5.67 (d,

Experimental part

^3J(H,H)=8.3 Hz, 1H, H-8), 2.88 (s, 3H, H-1), 2.75 (t, ^3J(H,H)=7.8 Hz, 2H, H-10), 2.33 (m, 2H, H-9), 1.39 (s, 6H, H-2, H-3) ppm.

^{13}C-NMR (101 MHz, CDCl$_3$, 25 °C): δ=174.8 (C-5), 154.1 (C-6), 141.4 (C$_{arom}$), 130.8 (C-7), 128.5 (C$_{arom}$), 128.4 (C$_{arom}$), 125.9 (C$_{arom}$), 117.1 (C-8), 61.2 (C-4), 34.8 (C-10), 29.9 (C-9), 24.5 (C-1), 22.2 (C-2, C-3) ppm.

EA Calcd: C, 70.5; H, 9.5; N, 10.3.
Found: C, 70.2; H, 9.3; N, 10.1.

MS (Ion trap, EI): *m/z* (%) = 273 (25, [M$^+$]), 181 (100), 153 (38), 72 (4), 56 (18).

Synthesis of 4,4-dimethyl-1-(4-phenyl-but-1-enyl)-piperidine-2,6-dione (3cc)

3cc was synthesized following method A using 3,3-dimethylglutarimide (141.2 mg, 1.00 mmol) and 4-phenyl-1-butyne (281 µL, 2.0 mmol) and purified by column chromatography (1:3 ethyl acetate/hexane), yielding the product as a colorless oil (214.1 mg, 88 %).

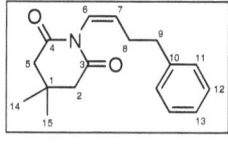

^{1}H-NMR (400 MHz, CDCl$_3$, 25 °C): δ=7.27 (t, ^3J(H,H)=7.3 Hz, 2H, H$_{arom}$), 7.17 (t, ^3J(H,H)=8.1 Hz, 3H, H$_{arom}$), 5.97 (d, ^3J(H,H)=8.3 Hz, 1H, H-6), 5.78 (d, ^3J(H,H)=8.1 Hz, 1H, H-7), 2.71 (t, ^3J(H,H)=7.8 Hz, 2H, H-9), 2.53 (s, 4H, H-2, 5), 2.17 (m, 2H, H-8), 1.10 (s, 6H, H-14, 15) ppm.

^{13}C-NMR (101 MHz, CDCl$_3$, 25 °C): δ=171.0 (C-3, 4), 141.4 (C$_{arom}$), 131.8 (C-6), 128.4 (C$_{arom}$), 126.0 (C$_{arom}$), 120.4 (C-7), 46.5 (C-2, 4), 34.4 (C-9), 29.5 (C-8), 29.2 (C-1), 27.8 (C-14, 15) ppm.

EA Calcd: C, 75.5; H, 7.7; N, 5.2.
Found: C, 75.1; H, 7.8; N, 5.0.

MS (Ion trap, EI): *m/z* (%) = 271 (5, [M$^+$]), 180 (100), 130 (43), 92 (21), 83 (48), 56 (42).

Experimental part

Synthesis of N-[(Z)-4-phenylbut-1-en-1-yl]glutarimide (3bc)

3bc was synthesized following method A using glutarimide (113.1 mg, 1.00 mmol) and 4-phenyl-1-butyne (281 µL, 2.0 mmol) and purified by column chromatography (1:3 ethyl acetate/hexane), yielding the product as a colorless oil (214.1 mg, 88 %).

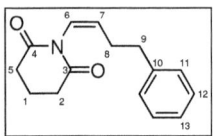

^{1}H-NMR	(400 MHz, CDCl$_3$, 25 °C): δ=7.28 (t, ^3J(H,H)=7.3 Hz, 2H, H$_{arom}$), 7.18 (t, ^3J(H,H)=7.8 Hz, 3H, H$_{arom}$), 5.97 (d, ^3J(H,H)=8.3 Hz, 1H, H-6), 5.78 (d, ^3J(H,H)=8.1 Hz, 1H, H-7), 2.70 (t, ^3J(H,H)=7.8 Hz, 2H, H-9), 2.65 (t, ^3J(H,H)=6.5 Hz, 4H, H-2, H-5), 2.20 (m, 2H, H-8), 1.92 (m, 2H, H-1) ppm.
^{13}C-NMR	(101 MHz, CDCl$_3$, 25 °C): δ=171.6 (C-3, C-4), 141.4 (C$_{arom}$), 131.5 (C-6), 128.4 (C$_{arom}$), 126.0 (C$_{arom}$), 120.6 (C-7), 34.4 (C-9), 32.9 (C-8), 29.3 (C-2, C-5), 17.2 (C-9) ppm.
EA	Calcd: C, 61.7; H, 7.0; N, 5.8. Found: C, 61.4; H, 6.8; N, 5.7.
MS	(Ion trap, EI): m/z (%) = 244 (10, [M$^+$]), 152 (100), 130 (55), 91 (24), 56 (55).

Synthesis of N-[(E)-hex-1-enyl]succinimide (4aa)

4aa was synthesized following method B using succinimide (99.1 mg, 1.00 mmol) and 1-hexyne (229 µL, 2.0 mmol) and purified by column chromatography (1:3 ethyl acetate/hexane), yielding the product as a yellowish oil (135.9 mg, 75%).

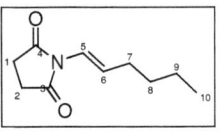

^{1}H-NMR	(400 MHz, CDCl$_3$, 25 °C): δ=6.52 (d, ^3J(H,H)=14.7 Hz, 1H, H-5), 6.36 (dt, ^3J(H,H)=15.0 Hz, 14.7 Hz, 1H, H-6), 2.66 (s, 4H, H-1, H-2), 2.05 (dt, ^3J(H,H)=14.6, 7.2, 2H, H-7), 1.24 – 1.37 (m, 4H, H-8, H-9), 0.83 (t, ^3J(H,H)=7.2 Hz, 3H, H-10) ppm.
^{13}C-NMR	(101 MHz, CDCl$_3$, 25 °C): δ=175.4 (C-3, C-4), 124.5 (C-5), 117.8 (C-6), 31.2 (C-7), 30.5 (C-8), 27.6 (C-1, C-2), 22.0 (C-9), 13.7 (C-10) ppm.
EA	Calcd: C, 66.2; H, 8.3; N, 7.7. Found: C, 66.4; H, 8.5; N, 7.9.
MS	(Ion trap, EI): m/z (%) = 182 (30, [M$^+$]), 138 (100), 100 (75), 82 (94), 56 (53).

Experimental part

Synthesis of N-[(E)-hex-1-en-1-yl]glutarimide (4ba)

4ba was synthesized following method B using glutarimide (113.1 mg, 1.00 mmol) and 1-hexyne (229 µL, 2.0 mmol) and purified by column chromatography (1:3 ethyl acetate/hexane), yielding the product as a colorless oil (78.1 mg, 40%).

^{1}H-NMR (400 MHz, CDCl$_3$, 25 °C): δ=6.18 (d, ^3J(H,H)=14.5 Hz, 1H, H-6), 5.94 (dt, ^3J(H,H)=14.5 Hz, 1H, H-7), 2.67 (t, ^3J(H,H)=7.5 Hz, 4H, H-2, H-5), 2.13 (dd, ^3J(H,H)=7.3 Hz, 2H, H-1), 1.91 – 1.96 (m, 2H, H-8) 1.37 – 1.43 (m, 4H, H-9, H-10), 0.87 (t, ^3J(H,H)=7.3 Hz, 3H, H-11) ppm.

^{13}C-NMR (101 MHz, CDCl$_3$, 25 °C): δ=172 (C-3, C-4), 131.5 (C-6), 119.7 (C-7), 32.2 (C-2, C-5), 31.1 (C-8), 30.3 (C-9), 22.1 (C-10), 16.9 (C-1), 13.8 (C-11) ppm.

EA Calcd: C, 67.7; H, 8.8; N, 7.2.
Found: C, 66.7; H, 8.8; N, 7.5.

MS (Ion trap, EI): *m/z* (%) = 195 (42, [M$^+$]), 138 (57), 114 (100), 86 (44), 56 (60).

Synthesis of N-[(E)-hex-1-en-1-yl]-3,3-dimethylglutarimide (4ca)

4ca was synthesized following method B using 3,3-dimethylglutarimide (141.2 mg, 1.00 mmol) and 1-hexyne (229 µL, 2.0 mmol) and purified by column chromatography (1:3 ethyl acetate/hexane), yielding the product as a yellowish oil (62.5 mg, 28%).

^{1}H-NMR (400 MHz, CDCl$_3$, 25 °C): δ=6.20 (d, ^3J(H,H)=14.5 Hz, 1H, H-6), 5.96 (dt, ^3J(H,H)=14.5 Hz, H-7), 2.52 (s, 4H, H-2, H-5), 2.11 – 2.15(m, 2H, H-8) 1.37 – 1.43 (m, 2H, H-9), 1.30 – 1.36 (m, 2H, H-10), 1.07 (s, 6H, H-12, H-13), 0.85 (t, ^3J(H,H)=7.2 Hz, 3H, H-11) ppm.

^{13}C-NMR (101 MHz, CDCl$_3$, 25 °C): δ=171.4 (C-3, C-4), 131.3 (C-6), 119.5 (C-7), 46.6 (C-2, C-5), 31.1 (C-8), 30.4 (C-9), 29.0 (C-1), 27.6 (C-12, C-13), 22.1 (C-10), 13.8 (C-11) ppm.

EA Calcd: C, 69.9; H, 9.5; N, 6.3.
Found: C, 69.5; H, 9.4; N, 6.4.

MS (Ion trap, EI): *m/z* (%) = 224 (83, [M$^+$]), 180 (66), 142 (100), 83 (82), 56 (46).

Experimental part

Synthesis of *N*-[(*E*)-2-phenylvinyl]succinimide (4ab) (CAS-N0: 120542-02-1)

4ab was synthesized following method B using succinimide (99.1 mg, 1.00 mmol) and phenylacetylene (220 μL, 2.0 mmol) and purified by column chromatography (1:3 ethyl acetate/hexane), yielding the product as a white solid (98.6 mg, 49 %).

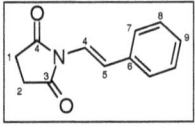

¹H-NMR (400 MHz, CDCl₃, 25 °C): δ=7.69 (d, ³*J*(H,H)=14.9 Hz, 1H, H-5), 7.45 (m, 2 H, H$_{arom}$), 7.35 (m, 2H, H$_{arom}$), 7.28 (m, 1 H, H$_{arom}$), 7.20 (d, ³*J*(H,H)=14.9 Hz, 1H, H-6), 2.81 (s, 4H, H-1, H-2) ppm.

¹³C-NMR (101 MHz, CDCl₃, 25 °C): δ=175.2 (C-3, C-4), 135.4 (C-5), 128.7 (C$_{arom}$), 127.9 (C$_{arom}$), 126.4 (C$_{arom}$), 121.9 (C$_{arom}$), 117.8 (C-6), 27.7 (C-1, C-2) ppm.

EA   Calcd: C, 71.6; H, 5.5; N, 7.0.
Found: C, 71.9; H, 5.6; N, 6.9.

MS   (Ion trap, EI): *m/z* (%) = 201 (100, [M⁺]), 172 (5), 119 (40), 89 (13), 55 (15).

The data correspond to those reported in the literature: T. Ogawa, T. Kiji, K. Hayami, H. Suzuki *Chem. Lett.* **1991**, *20*, 1443.

Synthesis of *N*-[(*E*)-4-phenylbut-1-en-1-yl]succinimide (4ac)

4ac was synthesized following method B using succinimide (99.1 mg, 1.00 mmol) and 4-phenyl-1-butyne (281 μL, 2.0 mmol) and purified by column chromatography (1:3 ethyl acetate/hexane), yielding the product as a white solid (178.8 mg, 78 %).

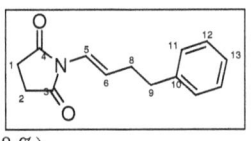

¹H-NMR (400 MHz, CDCl₃, 25 °C): δ=7.26 – 7.35 (m, 2H, H$_{arom}$), 7.22 (m, 3H H$_{arom}$), 6.69 (dt, ³*J*(H,H)=14.6 Hz, 1H, H-6), 6.49 (d, ³*J*(H,H)=14.9 Hz, 1H, H-5), 2.73 – 2.81 (m, 2H, H-9), 2.71 (s, 4H, H-1, H-2), 2.41 – 2.51 (m, 2H, H-8) ppm.

¹³C-NMR (101 MHz, CDCl₃, 25 °C): δ=175.3 (C-3, C-4), 141.2 (C$_{arom}$), 128.33 (C$_{arom}$), 128.27 (C$_{arom}$), 125.9 (C$_{arom}$), 123.5 (C-5), 118.4 (C-6), 35.6 (C-9), 32.8 (C-8), 27.7 (C-1, C-2) ppm.

EA   Calcd: C, 73.3; H, 6.6; N, 6.1.
Found: C, 73.5; H, 6.57; N, 5.8.

MS   (Ion trap, EI): *m/z* (%) = 229 (2, [M⁺]), 138 (100), 110 (15), 65 (18), 56 (27).

Experimental part

5.3 Addition of primary amides to alkynes

5.3.1 General procedure for the hydroamidation of terminal alkynes

Method A: An oven-dried flask was charged with the *N*-nucleophile (1.00 mmol), bis(2-methallyl)-cycloocta-1,5-diene-ruthenium(II) (16.0 mg, 0.05 mmol), 1,4-bis(dicyclohexylphosphino)butane (27.0 mg, 0.06 mmol) and ytterbium triflate (24.8 mg, 0.04 mmol) and flushed with nitrogen. Subsequently, dry DMF (3.0 mL), alkyne (2.00 mmol) and water (108 µL, 6.00 mmol) were added *via* syringe. The resulting solution was stirred for 6 h at 60 °C, then poured into an aqueous sodium bicarbonate solution (30 mL). The resulting mixture was extracted repeatedly with 20 mL portions of ethyl acetate, the combined organic layers were washed with water and brine, dried with magnesium sulfate, filtered, and the volatiles were removed in vacuo. The residue was purified by column chromatography (silica gel, ethyl acetate/hexane) to yield the *Z*-secondary enamide. The identity and purity of the products were confirmed by ^{1}H and ^{13}C NMR spectroscopy, mass spectroscopy and elemental analysis.

Method B: After the completion of the reaction following method A, 500 mg of 3 Å molecular sieves and 200 µL triethylamine were added into the reaction flask, and the reaction mixture was stirred for 24 h at 110 °C, then poured into an aqueous sodium bicarbonate solution (30 mL). The resulting mixture was extracted repeatedly with 20 mL portions of ethyl acetate, the combined organic layers were washed with water and brine, dried with magnesium sulfate, filtered, and the volatiles were removed in vacuo. The residue was purified by column chromatography (silica gel, ethyl acetate/hexane) to yield the *E*-secondary enamide. The identity and purity of the products were confirmed by ^{1}H and ^{13}C NMR spectroscopy, mass spectroscopy and elemental analysis.

Experimental part

5.3.2 Synthesis of the secondary enamides

Synthesis of *N*-((Z)-hex-1-en-1-yl)benzamide (9aa)

9aa was synthesized following method A using benzamide (121.1 mg, 1.00 mmol) and 1-hexyne (229 µL, 2.00 mmol) and purified by column chromatography (1:9 ethyl acetate/hexane), yielding the product as a yellowish oil (191.1 mg, 94%).

¹H-NMR (600 MHz, CDCl₃, 25 °C): δ=7.78 (d, ³*J*(H,H)=7.2 Hz, 2H, H$_{arom}$), 7.64 (d, ³*J*(H,H)=8.4 Hz, 1H, N-H), 7.52 (t, ³*J*(H,H)=7.4 Hz, 1H, H$_{arom}$), 7.45 (t, ³*J*(H,H)=7.6 Hz, 2H, H$_{arom}$), 6.90 (dd, ³*J*(H,H)=10.8, 9.2 Hz, 1H, H-6), 4.83-4.88 (m, 1H, H-7), 2.08 (qd, ³*J*(H,H)=7.3, 1.5 Hz, 2H, H-8), 1.40-1.44 (m, 2H, H-9), 1.34-1.39 (m, 2H, H-10), 0.92 (t, ³*J*(H,H)=7.2 Hz, 3H, H-11) ppm.

¹³C-NMR (151 MHz, CDCl₃, 25 °C): δ=164.3 (C-5), 134.0 (C$_{arom}$), 131.9 (C$_{arom}$), 128.7 (C$_{arom}$), 127.0 (C$_{arom}$), 121.1 (C-6), 112.3 (C-7), 31.4 (C-8), 25.5 (C-9), 22.3 (C-10), 14.0 (C-11) ppm.

EA   Calcd: C, 76.81; H, 8.43; N, 6.89.
      Found: C, 76.73; H, 8.49; N, 6.88.

MS   (Ion trap, EI): *m/z* (%) = 203 (8, [M⁺]), 160 (10), 122 (27), 105 (100), 77 (39), 51 (11).

Synthesis of *N*-[(Z)-2-phenyl-vinyl]benzamide (9ab) (CAS-N0: 5202-77-7)

9ab was synthesized following method A using benzamide (121.1 mg, 1.00 mmol) and 1-hexyne phenylacetylene (220 µL, 2.00 mmol) and purified by column chromatography (1:9 ethyl acetate/hexane), yielding the product as a white solid (216.6 mg, 97%).

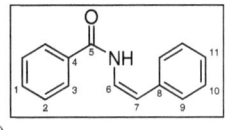

¹H-NMR (600 MHz, CDCl₃, 25 °C): δ=8.38 (d, ³*J*(H,H)=9.7 Hz, 1H, N-H), 7.75 (d, ³*J*(H,H)=7.2 Hz, 2H, H$_{arom}$), 7.53 (t, ³*J*(H,H)=7.4 Hz, 1H, H$_{arom}$), 7.44 (dt, ³*J*(H,H)=14.9, 7.5 Hz, 4H, H$_{arom}$), 7.35 (d, ³*J*(H,H)=7.4 Hz, 2H, H$_{arom}$), 7.28 (t, ³*J*(H,H)=7.3 Hz, 1H, H$_{arom}$), 7.20 (dd, ³*J*(H,H)=10.8, 9.7 Hz, 1H, H-6), 5.89 (d, ³*J*(H,H)=9.5 Hz, 1H, H-7) ppm.

Experimental part

13C-NMR (151 MHz, CDCl$_3$, 25 °C): δ=164.3 (C-5), 135.7 (C$_{arom}$), 133.3 (C$_{arom}$ 8), 132.1 (C$_{arom}$), 129.2 (C$_{arom}$), 128.8 (C$_{arom}$), 127.8 (C$_{arom}$), 127.0 (C$_{arom}$), 122.3 (C-6), 110.9 (C-7) ppm.

EA Calcd: C, 80.69; H, 5.87; N, 6.27.
Found: C, 80.36; H, 5.84; N, 6.06.

MS (Ion trap, EI): m/z (%) = 223 (54, [M$^+$]), 117 (5), 105 (100), 91 (5), 77 (46), 51 (14).

Synthesis of N-[(Z)-3,3-dimethyl-but-1-en-1-yl]benzamide (9ac)

9ac was synthesized following method A using benzamide (121.1 mg, 1.00 mmol) and 3,3-dimethyl-1-butyne (246 µL, 2.00 mmol) and purified by column chromatography (1:9 ethyl acetate/hexane), yielding the product as a white solid (185.0 mg, 91%).

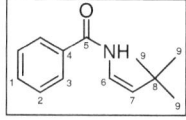

1H-NMR (600 MHz, CDCl$_3$, 25 °C): δ=8.02 (d, ^3J(H,H)=9.3 Hz, 1H, N-H), 7.75 (d, ^3J(H,H)=7.1 Hz, 2H, H$_{arom}$), 7.51 (t, ^3J(H,H)=7.4 Hz, 1H, H$_{arom}$), 7.44 (t, ^3J(H,H)=7.6 Hz, 2H, H$_{arom}$), 6.70-6.75 (m, 1H, H-6), 4.70 (d, ^3J(H,H)=10.0 Hz, 1H, H-7), 1.21 (s, 9H, H-9) ppm.

13C-NMR (151 MHz, CDCl$_3$, 25 °C): δ=163.7 (C-5), 133.8 (C$_{arom}$), 131.9 (C$_{arom}$), 128.7 (C$_{arom}$), 126.7 (C$_{arom}$), 121.5 (C-6), 118.7 (C-7), 32.6 (C-8), 30.9 (C-9) ppm.

EA Calcd C, 76.81; H, 8.43; N, 6.89.
Found: C, 76.49; H, 8.46; N, 6.72.

MS (Ion trap, EI): m/z (%) = 203 (5, [M$^+$]), 188 (49), 146 (7), 105 (100), 77 (40), 51 (11).

Experimental part

Synthesis of *N*-[(*Z*)-4-phenyl-but-1-en-1-yl]benzamide (9ad) (CAS-N0: 569340-56-3)

9ad was synthesized following method A using benzamide (121.1 mg, 1.00 mmol) and 4-phenyl-1-butyne (281 μL, 2.00 mmol) and purified by column chromatography (1:9 ethyl acetate/hexane), yielding the product as a white solid (231.2 mg, 92%)

^{1}H-NMR (600 MHz, CDCl$_3$, 25 °C): δ=7.61 (d, ^3J(H,H)=7.2 Hz, 2H, H$_{arom}$), 7.51 (t, ^3J(H,H)=7.4 Hz, 1H, H$_{arom}$), 7.41 (t, ^3J(H,H)=7.8 Hz, 2H, H$_{arom}$), 7.26-7.30 (m, 3H, H-11, N-H), 7.22-7.25 (m, 2H, H$_{arom}$), 7.16 (t, ^3J(H,H)=7.3 Hz, 1H, H$_{arom}$), 6.89 (dd, ^3J(H,H)=10.6, 9.1 Hz, 1H, H-6), 4.94 (q, ^3J(H,H)=7.9 Hz, 1H, H-7), 2.76 (t, ^3J(H,H)=7.3 Hz, 2H, H-9), 2.39-2.44 (m, 2H, H-8) ppm.

^{13}C-NMR (151 MHz, CDCl$_3$, 25 °C): δ=164.3 (C-5), 141.5 (C$_{arom}$), 133.7 (C$_{arom}$), 131.8 (C$_{arom}$), 128.53 (C$_{arom}$), 128.50 (C$_{arom}$), 128.4 (C$_{arom}$), 127.0 (C$_{arom}$), 126.2 (C$_{arom}$), 122.1 (C-6), 110.9 (C-7), 35.4 (C-9), 28.3 (C-8) ppm.

EA Calcd: C, 81.24; H, 6.82; N, 5.57.
Found: C, 81.37; H, 6.78; N, 5.69.

MS (Ion trap, EI): *m/z* (%) = 251 (4, [M$^+$]), 160 (47), 105 (100), 91 (27), 77 (37), 51 (11).

Synthesis of *N*-[(*Z*)-3-cyclohexyl-prop-1-en-1-yl]benzamide (9ae)

9ae was synthesized following method A using benzamide (121.1 mg, 1.00 mmol) and 3-cyclohexyl-1-propyne (289 μL, 2.00 mmol) and purified by column chromatography (1:9 ethyl acetate/hexane), yielding the product as a white solid (209.3 mg, 86%).

^{1}H-NMR (600 MHz, CDCl$_3$, 25 °C): δ=7.77 (d, ^3J(H,H)=7.4 Hz, 2H, H$_{arom}$), 7.67 (d, ^3J(H,H)=9.7 Hz, 1H, N-H), 7.51 (t, ^3J(H,H)=7.4 Hz, 1H, H$_{arom}$), 7.44 (t, ^3J(H,H)=7.6 Hz, 2H, H$_{arom}$), 6.93 (dd, ^3J(H,H)=10.7, 9.0 Hz, 1H, H-6), 4.87 (q, ^3J(H,H)=7.8 Hz, 1H, H-7), 1.93-1.97 (m, 2H, H-8), 1.73-1.77 (m, 2H, H$_{Cy}$), 1.70 (dt, ^3J(H,H)=13.0, 3.1 Hz, 2H, H$_{Cy}$), 1.63-1.67 (m, 1H, H$_{Cy}$), 1.36 (td,

^3J(H,H)=7.1, 3.8 Hz, 1H, H$_{Cy}$), 1.18-1.26 (m, 2H, H$_{Cy}$), 1.10-1.17 (m, 1H, H$_{Cy}$), 0.90-0.97 (m, 2H, H$_{Cy}$) ppm.

^{13}C-NMR (151 MHz, CDCl$_3$, 25 °C): δ=164.2 (C-5), 133.9 (C$_{arom}$), 131.8 (C$_{arom}$), 128.7 (C$_{arom}$), 127.0 (C$_{arom}$), 121.7 (C-6), 110.8 (C-7), 38.1 (C-8), 33.7 (C$_{Cy}$), 33.1 (C$_{Cy}$), 26.4 (C$_{Cy}$), 26.2 (C$_{Cy}$) ppm.

EA Calcd: C, 78.97; H, 8.70; N, 5.76.
Found: C, 78.95; H, 8.73; N, 5.55.

MS (Ion trap, EI): m/z (%) = 243 (17, [M$^+$]), 160 (23), 122 (26), 105 (100), 77 (31), 51 (7).

Synthesis of N-[(Z)-5-chloro-pent-1-en-1-yl]benzamide (9af)

9af was synthesized following method A using benzamide (121.1 mg, 1.00 mmol) and 5-chloro-1-pentyne (210 µL, 2.00 mmol) and purified by column chromatography (1:9 ethyl acetate/hexane), yielding the product as a white solid (203.6 mg, 91%).

^{1}H-NMR (600 MHz, CDCl$_3$, 25 °C): δ=7.95 (d, ^3J(H,H)=9.0 Hz, 1H, N-H), 7.80-7.84 (m, 2H, H$_{arom}$), 7.51 (t, ^3J(H,H)=7.4 Hz, 1H, H$_{arom}$), 7.44 (t, ^3J(H,H)=7.6 Hz, 2H, H$_{arom}$), 7.01 (dd, ^3J(H,H)=10.6, 9.2 Hz, 1H, H-6), 4.78 (q, ^3J(H,H)=8.3 Hz, 1H, H-7), 3.61-3.64 (m, 2H, H-10), 2.29-2.34 (m, 2H, H-9), 1.91 (dq, ^3J(H,H)=6.5, 6.3 Hz, 2H, H-8) ppm.

^{13}C-NMR (151 MHz, CDCl$_3$, 25 °C): δ=164.5 (C-5), 133.6 (C$_{arom}$), 131.9 (C$_{arom}$), 128.7 (C$_{arom}$), 127.1 (C$_{arom}$), 123.3 (C-6), 109.4 (C-7), 44.5 (C-10), 31.3 (C-9), 22.3 (C-8) ppm.

EA Calcd: C, 64.43; H, 6.31; N, 6.26.
Found: C, 64.10; H, 6.21; N, 5.98.

MS (Ion trap, EI): m/z (%) = 223 (18, [M$^+$]), 188 (7), 160 (7), 105 (100), 77 (29), 51 (10).

Experimental part

Synthesis of N-[(Z)-2-(4-fluoro-phenyl)-vinyl]benzamide (9ag)

9ag was synthesized following method A using benzamide (121.1 mg, 1.00 mmol) and 1-ethynyl-4-fluorobenzene (229 µL, 2.00 mmol) and purified by column chromatography (1:9 ethyl acetate/hexane), yielding the product as a white solid (238.9 mg, 99%).

1H-NMR (600 MHz, CDCl$_3$, 25 °C): δ=8.24 (d, ^3J(H,H)=10.0 Hz, 1H, N-H), 7.72-7.75 (m, 2H, H$_{arom}$), 7.51-7.55 (m, 1H, H$_{arom}$), 7.45 (t, ^3J(H,H)=7.7 Hz, 2H, H$_{arom}$), 7.29-7.33 (m, 2H, H$_{arom}$), 7.17 (dd, ^3J(H,H)=11.0, 9.7 Hz, 1H, H-6), 7.09-7.13 (m, 2H, H$_{arom}$), 5.84 (d, ^3J(H,H)=9.5 Hz, 1H, H-7) ppm.

13C-NMR (151 MHz, CDCl$_3$, 25 °C): δ=163.3 (d, ^2J=301 Hz, C-F), 160.7 (C-5), 133.2 (C-4), 132.2 (C-1), 131.7 (d, ^5J=2.8 Hz, C-8), 129.5 (d, ^4J=8.4 Hz, C-9), 128.8 (C-3), 127.0 (C-2), 122.4 (C-6), 116.2 (d, ^3J=22.2 Hz, C-10), 109.9 (C-7) ppm.

EA Calcd: C, 74.68; H, 5.01; N, 5.81.
Found: C, 74.42; H, 4.98; N, 5.68.

MS (Ion trap, EI): m/z (%) = 241 (56, [M$^+$]), 135 (3), 109 (6), 105 (100), 77 (37), 51 (12).

Synthesis of N-[(Z)-2-(4-methoxyphenyl)vinyl]benzamide (9ah) (CAS-N0: 128737-84-8)

9ah was synthesized following method A using benzamide (121.1 mg, 1.00 mmol) and 4-methoxy-phenylacetylene (259 µL, 2.00 mmol) and purified by column chromatography (1:8 ethyl acetate/hexane), yielding the product as a white solid (210.3 mg, 83%).

1H-NMR (600 MHz, CDCl3, 25 °C): δ=8.30 (d, ^3J(H,H)=10.2 Hz, 1H, N-H), 7.75 (d, ^3J(H,H)=7.4 Hz, 2H, H$_{arom}$), 7.52 (t, ^3J(H,H)=7.3 Hz, 1H, H$_{arom}$), 7.44 (t, ^3J(H,H)=7.7 Hz, 2H, H$_{arom}$), 7.27 (d, ^3J(H,H)=8.4 Hz, 2H, H$_{arom}$), 7.12 (t, ^3J(H,H)=10.2 Hz, 1H, H-6), 6.95 (d, ^3J(H,H)=8.4 Hz, 2H, H$_{arom}$), 5.83 (d, ^3J(H,H)=9.2 Hz, 1H, H-7), 3.82 (s, 3H, H-12) ppm.

13C-NMR (151 MHz, CDCl3, 25 °C): δ=164.2 (C-5), 158.5 (C$_{arom}$), 133.4 (C$_{arom}$), 132.0 (C$_{arom}$), 129.0 (C$_{arom}$), 128.8 (C$_{arom}$), 128.0 (C$_{arom}$), 127.0 (C$_{arom}$), 121.3 (C-6), 114.6, 110.7 (C-7), 55.3 (C-12) ppm.

- 152 -

Experimental part

EA Calcd: C, 75.87; H, 5.97; N, 5.53.
Found: C, 75.59; H, 5.82; N, 5.31.

MS (Ion trap, EI): m/z (%) = 253 (64, [M+]), 150 (12), 133 (8), 105 (100), 77 (45), 51 (15).

Synthesis of 1,1-dimethyl-3-[(Z)-2-phenylvinyl]urea (9bb)

9bb was synthesized following method A using 1,1-dimethylurea (88.1 mg, 1.00 mmol) and phenylacetylene (220 μL, 2.00 mmol) and purified by column chromatography (1:9 ethyl acetate/hexane), yielding the product as a white solid (87.6 mg, 46%).

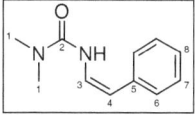

^{1}H-NMR (600 MHz, CDCl$_3$, 25 °C): δ=7.37 (t, ^3J(H,H)=7.8 Hz, 2H, H$_{arom}$), 7.29 (d, ^3J(H,H)=6.9 Hz, 2H, H$_{arom}$), 7.21 (s, 1H, H-8), 7.08 (s, 1H, N-H), 6.97 (d, ^3J(H,H)=9.7 Hz, 1H, H-3), 5.57 (d, ^3J(H,H)= 9.5 Hz, 1H, H-4), 2.95 (s, 6H, H-1) ppm.

^{13}C-NMR (151 MHz, CDCl$_3$, 25 °C): δ=154.8 (C-2), 136.7 (C$_{arom}$), 129.1 (C$_{arom}$), 127.5 (C$_{arom}$), 126.3 (C$_{arom}$), 124.7 (C-3), 106.0 (C-4), 36.2 (C-1) ppm.

EA Calcd: C, 69.45; H, 7.42; N, 14.72.
Found: C, 69.56; H, 7.13; N, 14.75.

MS (Ion trap, EI): *m/z* (%) = 190 (58, [M$^+$]), 145 (14), 117 (10), 89 (10), 72 (100), 44 (7).

Synthesis of 2,2-dimethyl-*N*-[(Z)-2-phenylvinyl]propanamide (9cb)
(CAS-N0: 189809-15-2)

9cb was synthesized following method A using 2,2-dimethyl-propionamide (101.2 mg, 1.00 mmol) and phenylacetylene (220 μL, 2.00 mmol) and purified by column chromatography (1:9 ethyl acetate/hexane), yielding the product as a white solid (178.9 mg, 88%).

- 153 -

Experimental part

¹H-NMR (400 MHz, CDCl3, 25 °C): δ=7.92 (s, 1H, N-H), 7.35-7.45 (m, 2H, H$_{arom}$), 7.21-7.31 (m, 3H, H$_{arom}$), 6.99 (dd, 3J(H,H, H-3)=10.9, 9.7 Hz, 1H), 5.75 (d, 3J(H,H)=9.6 Hz, 1H, H-4), 1.22 (s, 9H, H-1) ppm.

¹³C-NMR (101 MHz, CDCl3, 25 °C): δ=175.6 (C-3), 135.8 (C$_{arom}$), 129.1 (C$_{arom}$), 127.6 (C$_{arom}$), 126.8 (C$_{arom}$), 122.4 (C-4), 109.7 (C-5), 38.8 (C-2), 27.2 (C-1) ppm.

EA Calcd: C, 76.81; H, 8.43; N, 6.89.
Found: C, 76.81; H, 8.49; N, 6.75.

MS (Ion trap, EI): m/z (%) = 203 (100, [M+]), 160 (3), 119 (32), 91 (6), 65 (3), 57 (29).

Synthesis of 2-methyl-N-[(Z)-2-phenylvinyl]acrylamide (9db)

9db was synthesized following method A using 2-methyl-acrylamide (85.1 mg, 1.00 mmol) and phenylacetylene (220 µL, 2.00 mmol) and purified by column chromatography (1:9 ethyl acetate/hexane), yielding the product as a yellowish oil (168.5 mg, 90%).

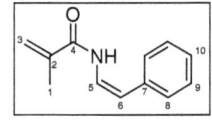

¹H-NMR (600 MHz, CDCl$_3$, 25 °C): δ=8.06 (d, ^3J(H,H)=7.4 Hz, 1H, N-H), 7.39 (t, ^3J(H,H)=7.8 Hz, 2H, H$_{arom}$), 7.28 (d, ^3J(H,H),=7.2 Hz, 2H, H$_{arom}$), 7.24 (t, ^3J(H,H)=7.4 Hz, 1H, H$_{arom}$), 7.03 (dd, ^3J(H,H)=11.1, 9.6 Hz, 1H, H-5), 5.80 (d, ^3J(H,H)=9.7 Hz, 1H, H-6), 5.72 (s, 1H, H-3), 5.43 (d, ^3J(H,H)=1.0 Hz, 1H, H-3), 1.97 (s, 3H, H-1) ppm.

¹³C-NMR (151 MHz, CDCl$_3$, 25 °C): δ=165.1 (C-4), 139.2 (C-2), 135.7 (C$_{arom}$), 129.1 (C$_{arom}$), 127.7 (C$_{arom}$), 126.9 (C$_{arom}$), 122.1 (C-5), 121.0 (C-3), 110.5 (C-6), 18.4 (C-1) ppm.

EA Calcd: C, 76.98; H, 7.00; N, 7.48.
Found: C, 76.81; H, 7.02; N, 7.15.

MS (Ion trap, EI): m/z (%) = 187 (100, [M⁺]), 159 (85), 144 (25), 117 (24), 69 (77), 41 (58).

Experimental part

Synthesis of 2-acetamido-*N*-[(Z)-2-phenylvinyl]acetamide (9eb)

9eb was synthesized following method A using 2-acetamidoacetamide (116.1 mg, 1.00 mmol) and phenylacetylene (220 µL, 2.00 mmol) and purified by column chromatography (1:9 ethyl acetate/hexane), yielding the product as a white solid (209.5 mg, 96%).

¹H-NMR (400 MHz, CDCl₃, 25 °C): δ=8.49 (d, ³*J*(H,H)=10.6 Hz, 1H, N-H), 7.36 (t, ³*J*(H,H)=7.5Hz, 2H, H_arom), 7.21-7.28 (m, 3H, H_arom), 6.83 (dd, ³*J*(H,H)=11.1, 9.7 Hz, 1H, H-5), 6.62 (s, 1H, N-H), 5.76 (d, ³*J*(H,H)=9.9 Hz, 1H, H-6), 3.96 (s, 2H, H-3), 1.96 (s, 3H, H-1) ppm.

¹³C-NMR (101 MHz, CDCl₃, 25 °C): δ=170.9 (C-2), 166.9 (C-4), 135.1 (C_arom), 128.9 (C_arom), 128.0 (C_arom), 127.1 (C_arom), 121.0 (C-5), 111.4 (C-6), 43.6 (C-3), 22.7 (C-1) ppm.

EA Calcd: C, 66.04; H, 6.47; N, 12.84.
Found: C, 66.32; H, 6.57; N, 12.61.

MS (Ion trap, EI): *m/z* (%) = 218 (9, [*M*⁺]), 119 (100), 104 (4), 91 (10), 65 (4), 43 (11).

Synthesis of *N*-[(Z)-2-phenylvinyl]butyramide (9fb)

9fb was synthesized following method A using butyramide (87.1 mg, 1.00 mmol) and phenylacetylene (220 µL, 2.00 mmol) and purified by column chromatography (1:9 ethyl acetate/hexane), yielding the product as a yellowish oil (160.9 mg, 85%).

¹H-NMR (600 MHz, CDCl₃, 25 °C): δ=7.61 (d, ³*J*(H,H)=7.9 Hz, 1H, N-H), 7.39 (t, ³*J*(H,H)=7.7 Hz, 2H, H_arom), 7.23-7.28 (m, 3H, H_arom), 6.98 (dd, ³*J*(H,H)=11.0, 9.7 Hz, 1H, H-5), 5.73 (d, ³*J*(H,H)=9.5 Hz, 1H, H-6), 2.22 (t, ³*J*(H,H)=7.4 Hz, 2H, H-3), 1.65-1.72 (m, 2H, H-2), 0.97 (t, ³*J*(H,H)=7.3 Hz, 3H, H-1) ppm.

¹³C-NMR (151 MHz, CDCl₃, 25 °C): δ=170.4 (C-4), 135.7 (C_arom), 129.0 (C_arom), 127.8 (C_arom), 126.8 (C_arom), 122.0 (C-5), 109.5 (C-6), 38.6 (C-3), 18.7 (C-2), 13.7 (C-1) ppm.

Experimental part

EA Calcd: C, 76.16; H, 7.99; N, 7.40.
Found: C, 75.83; H, 7.92; N, 6.89.

MS (Ion trap, EI): m/z (%) = 189 (50, [M$^+$]), 119 (100), 104 (2), 65 (4), 51 (2), 41 (8).

Synthesis of 3-nitro-*N*-[(Z)-2-phenylvinyl]benzamide (9gb) (CAS-N0: 389572-33-2)

9gb was synthesized following method A using 3-nitrobenzamide (168.2 mg, 1.00 mmol) and phenylacetylene (220 µL, 2.00 mmol) and purified by column chromatography (1:9 ethyl acetate/hexane), yielding the product as a white solid (214.6 mg, 80%).

^{1}H-NMR (600 MHz, CDCl$_3$, 25 °C): δ=8.58 (d, ^3J(H,H)=10.8 Hz, 1H, N-H), 8.54 (s, 1H, H$_{arom}$), 8.30 (d, ^3J(H,H)=8.2 Hz, 1H, H$_{arom}$), 8.04 (d, ^3J(H,H)=7.9 Hz, 1H, H$_{arom}$), 7.62 (t, ^3J(H,H)=7.9 Hz, 1H, H$_{arom}$), 7.37 (t, ^3J(H,H)=7.9 Hz, 2H, H$_{arom}$), 7.32 (d, ^3J(H,H)=7.2 Hz, 2H, H$_{arom}$), 7.22 (t, ^3J(H,H)=7.4 Hz, 1H, H$_{arom}$), 7.08 (t, ^3J(H,H)=10.5 Hz, 1H, H-8), 5.92 (d, ^3J(H,H)=9.5 Hz, 1H, H-9) ppm.

^{13}C-NMR (151 MHz, CDCl$_3$, 25 °C): δ=162.1 (C-7), 148.0 (C$_{arom}$), 135.1 (C$_{arom}$), 134.8 (C$_{arom}$), 132.7 (C$_{arom}$), 129.8 (C$_{arom}$), 129.1 (C$_{arom}$), 127.8 (C$_{arom}$), 127.2 (C$_{arom}$), 126.3 (C$_{arom}$), 122.2 (C$_{arom}$), 121.7 (C-8), 112.5 (C-9) ppm.

EA Calcd: C, 67.16; H, 4.51; N, 10.44.
Found: C, 67.15; H, 4.68; N, 10.52.

MS (Ion trap, EI): m/z (%) = 268 (64, [M$^+$]), 150 (100), 120 (20), 104 (36), 76 (25), 50 (13).

Synthesis of 4-fluoro-*N*-[(Z)-2-phenylvinyl]benzamide (9hb)

9hb was synthesized following method A using 4-fluorobenzamide (139.1 mg, 1.00 mmol) and phenylacetylene (220 µL, 2.00 mmol) and purified by column chromatography (1:9 ethyl acetate/hexane), yielding the product as a white solid (224.4 mg, 93%).

¹H-NMR	(400 MHz, CDCl3, 25 °C): δ=8.24 (d, ^3J(H,H)=10.6 Hz, 1H, N-H), 7.63-7.69 (m, 2H, H$_{arom}$), 7.33 (t, ^3J(H,H)=7.7 Hz, 2H, H$_{arom}$), 7.22-7.27 (m, 2H, H$_{arom}$), 7.18 (t, ^3J(H,H)=7.7 Hz, 1H, H-6), 6.99-7.09 (m, 3H), 5.79 (d, ^3J(H,H)=9.5 Hz, 1H, H-7) ppm.
¹³C-NMR	(101 MHz, CDCl3, 25 °C): δ=165.0 (d, ^2J=253.4 Hz, C-1), 163.2 (C-5), 135.7 (C$_{arom}$), 129.6 (d, ^5J=2.8 Hz, C-4), 129.4 (d, ^4J=9.3 Hz, C-3), 129.2 (C$_{arom}$), 127.8 (C$_{arom}$), 127.1 (C$_{arom}$), 122.3 (C-6), 115.8 (d, ^3J=22.2 Hz, C-2), 111.0 (C-7) ppm.
EA	Calcd: C, 74.68; H, 5.01; N, 5.81. Found: C, 74.67; H, 5.11; N, 5.66.
MS	(Ion trap, EI): m/z (%) = 242 (100, [M+1]⁺), 241 (52, [M+]), 123 (100), 95 (21), 75 (7), 50 (3).

Synthesis of 4-methoxy-*N*-[(Z)-2-phenylvinyl]benzamide (9ib) (CAS-N0: 159824-57-4)

9ib was synthesized following method A using 4-methoxybenzamide (151.2 mg, 1.00 mmol) and phenylacetylene (220 μL, 2.00 mmol) and purified by column chromatography (1:8 ethyl acetate/hexane), yielding the product as a white solid (243.2 mg, 96%).

¹H-NMR	(600 MHz, CDCl₃, 25 °C): δ=8.31 (d, ^3J(H,H)=10.5 Hz, 1H, N-H), 7.70-7.73 (m, 2H, H$_{arom}$), 7.42 (t, ^3J(H,H)=7.8 Hz, 2H, H$_{arom}$), 7.34 (d, ^3J(H,H)=7.2 Hz, 2H, H$_{arom}$), 7.27 (t, ^3J(H,H)=7.3 Hz, 1H, H$_{arom}$), 7.16-7.21 (m, 1H, H-7), 6.91-6.94 (m, 2H, H$_{arom}$), 5.85 (d, ^3J(H,H)=9.5 Hz, 1H, H-8), 3.84 (s, 3H, H-1) ppm.
¹³C-NMR	(151 MHz, CDCl₃, 25 °C): δ=163.8 (C-6), 162.7 (C$_{arom}$), 135.9 (C$_{arom}$), 129.2 (C$_{arom}$), 129.0 (C$_{arom}$), 127.8 (C$_{arom}$), 126.9 (C$_{arom}$), 125.5 (C$_{arom}$), 122.5 (C-7), 114.0 (C$_{arom}$), 110.2 (C-8), 55.4 (C-1) ppm.
EA	Calcd: C, 75.87; H, 5.97; N, 5.53. Found: C, 75.97; H, 6.12; N, 5.43.
MS	(Ion trap, EI): m/z (%) = 253 (32, [M⁺]), 135 (100), 107 (9), 92 (8), 77 (16), 63 (6).

Experimental part

Synthesis of ethyl oxo{[(Z)-2-phenylvinyl]amino}acetate (9jb)

9jb was synthesized following method A using oxalamic acid ethyl ester (117.1 mg, 1.00 mmol) and phenylacetylene (220 µL, 2.00 mmol) but owing to the high hydrolytic instability of the product, no water was added. Purification by column chromatography (1:7 ethyl acetate/hexane), yielded the product as a white solid (171.0 mg, 78%).

¹H-NMR (600 MHz, CDCl₃, 25 °C): δ=9.24 (d, ³J(H,H)=9.2 Hz, 1H, N-H), 7.41 (t, ³J(H,H)=7.7 Hz, 2H, H$_{arom}$), 7.27-7.32 (m, 3H, H$_{arom}$), 6.91 (dd, ³J(H,H)=11.8, 9.5 Hz, 1H, H-5), 5.98 (d, ³J(H,H)=9.5 Hz, 1H, H-6), 4.36 (q, ³J(H,H)=7.2 Hz, 2H, H-2), 1.38 (t, ³J(H,H)=7.2 Hz, 3H, H-1) ppm.

¹³C-NMR (151 MHz, CDCl₃, 25 °C): δ=160.2 (C-4), 153.6 (C-3), 134.7 (C$_{arom}$), 129.2 (C$_{arom}$), 127.9 (C$_{arom}$), 127.6 (C$_{arom}$), 120.3 (C-5), 114.2 (C-6), 63.6 (C-2), 13.9 (C-1) ppm.

EA Calcd: C, 65.74; H, 5.98; N, 6.39.
Found: C, 66.00; H, 5.99; N, 6.27.

MS (Ion trap, EI): m/z (%) = 219 (100, [M^+]), 146 (3), 118 (10), 117 (11), 91 (2), 89 (3).

Synthesis of 2-cyano-*N*-[(Z)-2-phenylvinyl]acetamide (9kb)

9kb was synthesized following method A using 2-cyano-acetamide (84.1 mg, 1.00 mmol) and phenylacetylene (220 µL, 2.00 mmol) and purified by column chromatography (1:6 ethyl acetate/hexane), yielding the product as a white solid (57.7 mg, 31%).

¹H-NMR (600 MHz, CDCl₃, 25 °C): δ=8.24 (s, 1H, N-H), 7.41 (t, ³J(H,H)=7.7 Hz, 2H, H$_{arom}$), 7.29 (s, 1H, H$_{arom}$), 7.28 (d, ³J(H,H)=4.1 Hz, 2H, H$_{arom}$), 6.86-6.90 (m, 1H, H-4), 5.93 (d, ³J(H,H)=9.5 Hz, 1H, H-5), 3.40 (s, 2H, H-2) ppm.

¹³C-NMR (151 MHz, CDCl₃, 25 °C): δ=158.4 (C-3), 134.5 (C$_{arom}$), 129.3 (C$_{arom}$), 127.8 (C$_{arom}$), 127.6 (C$_{arom}$), 120.7 (C-4), 113.9 (C-1), 113.1 (C-5), 25.9 (C-2) ppm.

EA Calcd: C, 70.95; H, 5.41; N, 15.04.

Experimental part

Found: C, 71.07; H, 5.12; N, 15.07.

MS   (Ion trap, EI): *m/z* (%) = 186 (100, [M⁺]), 145 (13), 118 (53), 91 (52), 77 (7), 65 (12).

Synthesis of ethyl 3-oxo-3{[(Z)-2-phenylvinyl]amino}propanoate (9lb)

9lb was synthesized following method A using malonamic acid ethyl ester (131.1 mg, 1.00 mmol) and phenylacetylene (220 µL, 2.00 mmol) but owing to the high hydrolytic instability of the product, no water was added. Purification by column chromatography (1:6 ethyl acetate/hexane), yielded the product as a white solid (165.6 mg, 71%).

^{1}H-NMR   (600 MHz, CDCl$_3$, 25 °C): δ=7.34 (t, ^3J(H,H)=7.6 Hz, 2H, H$_{arom}$), 7.28-7.31 (m, 2H, H$_{arom}$), 7.26 (t, ^3J(H,H)=7.9 Hz, 1H, H$_{arom}$), 6.78 (d, ^3J(H,H)=11.3 Hz, 1H, N-H), 6.50 (s, 1H, H-4), 6.35 (s, 1H, H-4), 5.95 (t, ^3J(H,H)=10.9 Hz, 1H, H-6), 4.43 (d, ^3J(H,H)=10.5 Hz, 1H, H-7), 4.16-4.22 (m, 2H, H-2), 1.22-1.28 (m, 3H, H-1) ppm.

^{13}C-NMR   (151 MHz, CDCl$_3$, 25 °C): δ=169.6 (C-3), 169.5 (C-5), 135.4 (C-7), 134.1 (C$_{arom}$), 128.5 (C$_{arom}$), 128.4 (C$_{arom}$), 127.6 (C$_{arom}$), 123.3 (C-6), 61.9 (C-2), 52.4 (C-4), 13.9 (C-1) ppm.

EA   Calcd: C, 66.94; H, 6.48; N, 6.00.
Found: C, 66.57; H, 6.17; N, 5.77.

MS   (Ion trap, EI): *m/z* (%) = 233 (6, [M⁺]), 204 (100), 170 (41), 102 (72), 77 (12), 50 (25).

Synthesis of (*E*)-3-phenyl-*N*-[(Z)-2-phenylvinyl]acrylamide (Lansiumamide A) (9mb) (CAS-N0: 121817-36-5)

9mb was synthesized following method A using 3-phenyl-acrylamide (147.2 mg, 1.00 mmol) and phenylacetylene (220 µL, 2.00 mmol) and purified by column chromatography (1:8 ethyl acetate/hexane), yielding the product as a white solid (244.3 mg, 98%).

^{1}H-NMR   (600 MHz, CDCl$_3$, 25 °C): δ=7.87 (d, ^3J(H,H)=11.0 Hz, 1H, N-H), 7.71 (d, ^3J(H,H)=15.6 Hz, 1H, H-5), 7.50 (dd, ^3J(H,H)=6.5, 2.7 Hz, 2H,H$_{arom}$), 7.41 (t,

^3J(H,H)=7.7 Hz, 2H,H$_{arom}$), 7.31-7.37 (m, 5H,H$_{arom}$), 7.27 (t, ^3J(H,H)=7.4 Hz, 1H,H$_{arom}$), 7.12 (dd, ^3J(H,H)=11.3, 9.7 Hz, 1H, H-8), 6.40 (d, ^3J(H,H)=15.6 Hz, 1H, H-6), 5.82 (d, ^3J(H,H)=9.5 Hz, 1H, H-9) ppm.

^{13}C-NMR (151 MHz, CDCl$_3$, 25 °C): δ=163.1 (C-7), 143.0 (C-5), 135.7 (C$_{arom}$), 134.4 (C$_{arom}$), 130.1 (C$_{arom}$), 129.1 (C$_{arom}$), 128.8 (C$_{arom}$), 128.0 (C$_{arom}$), 127.9 (C$_{arom}$), 127.0 (C$_{arom}$), 122.2 (C-8), 119.4 (C-6), 110.6 (C-9) ppm.

EA Anal. Calcd for C$_{17}$H$_{15}$NO: C, 81.90; H, 6.06; N, 5.62. Found: C, 81.70; H, 6.05; N, 5.50.

MS (Ion trap, EI): m/z (%) = 249 (44, [M$^+$]), 131 (100), 119 (35) 103 (50), 77 (23), 51 (10).

Synthesis of 2-bromo-4,5-dimethoxy-*N*-[(*Z*)-2-(4-methoxy-phenyl)vinyl]benzamide (9nh)
(CAS-N0: 128580-95-0)

9nh was synthesized following method A using 2-bromo-4,5-dimethoxy-benzamide (260.1 mg, 1.00 mmol) and 4-methoxy-phenylacetylene (259 μL, 2.00 mmol) and purified by column chromatography (1:6 ethyl acetate/hexane), yielding the product as a white solid (235.4 mg, 60%).

^{1}H-NMR (400 MHz, CDCl$_3$, 25 °C): δ=8.53 (d, ^3J(H,H)=10.9 Hz, 1H, N-H), 7.33 (s, 1H, H$_{arom}$), 7.28-7.32 (m, 2H, H$_{arom}$), 7.08 (dd, ^3J(H,H)=11.1, 9.7 Hz, 1H, H-10), 6.97 (s, 1H, H$_{arom}$), 6.88-6.92 (m, 2H, H$_{arom}$), 5.86 (d, ^3J(H,H)=9.5 Hz, 1H, H-11), 3.88 (2 signals, 2 singulets, each 3H, H-1 and H-2), 3.80 (s, 3H, H-16) ppm.

^{13}C-NMR (101 MHz, CDCl$_3$, 25 °C): δ=163.7 (C-9), 158.6 (C$_{arom}$), 151.5 (C$_{arom}$), 148.6 (C$_{arom}$), 129.3 (C$_{arom}$), 127.8 (C$_{arom}$), 127.7 (C$_{arom}$), 121.0 (C-10), 116.0 (C$_{arom}$), 114.4 (C$_{arom}$), 113.9 (C$_{arom}$), 111.3 (C$_{arom}$), 109.8 (C-11), 56.3 and 56.2 (C-1 and C-2), 55.3 (C-16) ppm.

MS (Ion trap, EI): m/z (%) = 391 (49, [M$^+$]), 393 (44, [*M*$^+$]), 245 (100), 132 (15), 104 (12), 77 (5).

Experimental part

Synthesis of *N*-((*E*)-hex-1-en-1-yl)benzamide (10aa) (CAS-N0: 474352-83-5)

10aa was synthesized following method B using benzamide (121.1 mg, 1.00 mmol) and 1-hexyne (229 µL, 2.00 mmol) and purified by column chromatography (1:9 ethyl acetate/hexane), yielding the product as a white solid (180.9 mg, 89%).

^{1}H-NMR (600 MHz, CDCl$_3$, 25 °C): δ=7.78 (d, ^3J(H,H)=7.4 Hz, 2H, H$_{arom}$), 7.69 (d, ^3J(H,H)=8.4 Hz, 1H, N-H), 7.50 (t, ^3J(H,H)=7.3 Hz, 1H, H$_{arom}$), 7.43 (t, ^3J(H,H)=7.7 Hz, 2H, H$_{arom}$), 6.94 (dd, ^3J(H,H)=14.0, 10.6 Hz, 1H, H-6), 5.30 (dt, ^3J(H,H)=14.3, 7.2 Hz, 1H, H-7), 2.07 (q, ^3J(H,H)=6.8 Hz, 2H, H-8), 1.30-1.38 (m, 4H, H-9, 10), 0.89 (t, ^3J(H,H)=7.2 Hz, 3H, H-11) ppm.

^{13}C-NMR (151 MHz, CDCl$_3$, 25 °C): δ=164.3 (C-5), 133.9 (C$_{arom}$), 131.9 (C$_{arom}$), 128.7 (C$_{arom}$), 127.0 (C$_{arom}$), 121.1 (C-6), 112.3 (C-7), 31.4 (C-8), 25.5 (C-9), 22.3 (C-10), 13.9 (C-11) ppm.

MS (Ion trap, EI): *m/z* (%) = 203 (17, [M$^+$]), 122 (26), 105 (100), 77 (41), 51 (12), 44 (15).

Synthesis of *N*-[(*E*)-2-phenyl-vinyl]benzamide (10ab) (CAS-N0: 78007-47-3)

10ab was synthesized following method B using benzamide (121.1 mg, 1.00 mmol) and phenylacetylene (220 µL, 2.00 mmol) and purified by column chromatography (1:8 ethyl acetate/hexane), yielding the product as a white solid (205.4 mg, 92%).

^{1}H-NMR (600 MHz, CDCl$_3$, 25 °C): δ=8.06 (d, ^3J(H,H)=10.0 Hz, 1H, N-H), 7.84-7.86 (m, 2H, H$_{arom}$), 7.73 (dd, ^3J(H,H)=14.6, 10.8 Hz, 1H, H-6), 7.52-7.56 (m, 1H, H$_{arom}$), 7.46 (t, ^3J(H,H)=7.7 Hz, 2H, H$_{arom}$), 7.33-7.36 (m, 2H, H$_{arom}$), 7.29 (t, ^3J(H,H)=7.7 Hz, 2H, H$_{arom}$), 7.19 (t, ^3J(H,H)=7.3 Hz, 1H, H$_{arom}$), 6.27 (d, ^3J(H,H)=14.6 Hz, 1H, H-7) ppm.

^{13}C-NMR (151 MHz, CDCl$_3$, 25 °C): δ=164.5 (C-5), 135.9 (C$_{arom}$), 133.4 (C$_{arom}$), 132.1 (C$_{arom}$), 128.8 (C$_{arom}$), 128.7 (C$_{arom}$), 127.1 (C$_{arom}$), 126.8 (C$_{arom}$), 125.6 (C$_{arom}$), 123.0 (C-6), 113.6 (C-7) ppm.

Experimental part

MS (Ion trap, EI): m/z (%) = 223 (49, [M$^+$]), 105 (100), 91 (7), 77 (53), 51 (16), 44 (11).

Synthesis of *N*-[(*E*)-3,3-dimethyl-but-1-en-1-yl]benzamide (10ac)

10ac was synthesized following method B using benzamide (121.1 mg, 1.00 mmol) and 3,3-dimethyl-1-butyne (246 µL, 2.00 mmol) and purified by column chromatography (1:9 ethyl acetate/hexane), yielding the product as a white solid (158.6 mg, 79%).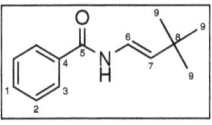

^{1}H-NMR (600 MHz, CDCl$_3$, 25 °C): δ=8.31 (d, ^3J(H,H)=10.0 Hz, 1H, N-H), 7.78-7.81 (m, 2H, H$_{arom}$), 7.44 (t, ^3J(H,H)=7.4 Hz, 1H, H$_{arom}$), 7.35 (t, ^3J(H,H)=7.7 Hz, 2H, H$_{arom}$), 6.88 (dd, ^3J(H,H)=14.6, 10.2 Hz, 1H, H-6), 5.40 (d, ^3J(H,H)=14.6 Hz, 1H, H7), 1.01 (s, 9H, H-9) ppm.

^{13}C-NMR (151 MHz, CDCl$_3$, 25 °C): δ=164.7 (C-5), 133.8 (C$_{arom}$), 131.5 (C$_{arom}$), 128.4 (C$_{arom}$), 127.0 (C$_{arom}$), 125.9 (C-6), 119.4 (C-7), 31.9 (C8-8), 29.8 (C-9) ppm.

EA Calcd: C, 76.81; H, 8.43; N, 6.89.
Found: C, 76.53; H, 8.46; N, 6.88.

MS (Ion trap, EI): m/z (%) = 203 (10, [M$^+$]), 172 (45), 139 (58), 105 (77), 77 (45), 44 (100).

Synthesis of *N*-[(*E*)-4-phenyl-but-1-en-1-yl]benzamide (10ad)

10ad was synthesized following method B using benzamide (121.1 mg, 1.00 mmol) and 4-phenyl-1-butyne (281 µL, 2.00 mmol) and purified by column chromatography (1:8 ethyl acetate/hexane), yielding the product as a white solid (213.6 mg, 85%).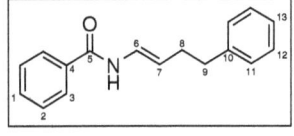

^{1}H-NMR (400 MHz, CDCl$_3$, 25 °C): δ=8.03 (d, ^3J(H,H)=10.2 Hz, 1H, N-H), 7.77-7.81 (m, 2H, H$_{arom}$), 7.46-7.52 (m, 1H, H$_{arom}$), 7.40 (t, ^3J(H,H)=7.5 Hz, 2H, H$_{arom}$), 7.28 (t, ^3J(H,H)=7.3 Hz, 2H, H$_{arom}$), 7.16-7.22 (m, 3H, H$_{arom}$), 7.00 (dd, ^3J(H,H)=14.3, 10.6 Hz, 1H, H-6), 5.35 (dt, ^3J(H,H)=14.3, 7.2 Hz, 1H, H-7), 2.67-2.73 (m, 2H, H-8), 2.39 (t, ^3J(H,H)=7.2 Hz, 2H, H-9) ppm.

Experimental part

13C-NMR (101 MHz, CDCl$_3$, 25 °C): δ=164.4 (C-5), 141.5 (C$_{arom}$), 133.8 (C$_{arom}$), 131.7 (C$_{arom}$), 128.6 (C$_{arom}$), 128.4 (C$_{arom}$), 128.3 (C$_{arom}$), 127.0 (C$_{arom}$), 125.9 (C$_{arom}$), 123.4 (C-6), 113.2 (C-7), 36.3 (C-8), 31.5 (C-9) ppm.

EA Calcd: C, 81.24; H, 6.82; N, 5.57.
Found: C, 80.87; H, 6.88; N, 5.28.

MS (Ion trap, EI): m/z (%) = 251 (12, [M$^+$]), 225 (25), 207 (52), 160 (47), 105 (100), 77 (36).

Synthesis of N-[(E)-3-cyclohexyl-prop-1-en-1-yl]benzamide (10ae)

10ae was synthesized following method B using benzamide (121.1 mg, 1.00 mmol) and 3-cyclohexyl-1-propyne (289 µL, 2.00 mmol) and purified by column chromatography (1:9 ethyl acetate/hexane), yielding the product as a white solid (206.8 mg, 85%).

1H-NMR (400 MHz, CDCl$_3$, 25 °C): δ=7.82 (d, ^3J(H,H)=9.9 Hz, 1H, N-H), 7.75-7.80 (m, 2H, H$_{arom}$), 7.46-7.51 (m, 1H, H$_{arom}$), 7.41 (t, ^3J(H,H)=7.5 Hz, 2H, H$_{arom}$), 6.91 (dd, ^3J(H,H)=14.3, 10.6 Hz, 1H, H-6), 5.29 (dt, ^3J(H,H)=14.5, 7.3 Hz, 1H, H-7), 1.94 (dd, ^3J(H,H)=7.2, 7.3 Hz, 2H, H-8), 1.62-1.72 (m, 6H, H$_{Cy}$), 1.15-1.26 (m, 3H, H$_{Cy}$), 0.83-0.94 (m, 2H, H$_{Cy}$) ppm.

13C-NMR (101 MHz, CDCl$_3$, 25 °C): δ=164.2 (C-5), 133.9 (C$_{arom}$), 131.7 (C$_{arom}$), 128.6 (C$_{arom}$), 127.0 (C$_{arom}$), 123.4 (C-6), 112.8 (C-7), 38.4 (C$_{Cy}$), 37.6 (C$_{Cy}$), 33.0 (C$_{Cy}$), 26.5 (C$_{Cy}$), 26.3 (C$_{Cy}$) ppm.

EA Calcd: C, 78.97; H, 8.70; N, 5.76.
Found: C, 78.93; H, 8.72; N, 5.78.

MS (Ion trap, EI): *m/z* (%) = 243 (12, [M$^+$]), 160 (23), 122 (27), 105 (100), 77 (33), 51 (9).

Experimental part

Synthesis of 2-methyl-N-[(E)-2-phenylvinyl]acrylamide (10db)

10db was synthesized following method B using 2-methyl-acrylamide (85.1 mg, 1.00 mmol) and phenylacetylene (220 µL, 2.00 mmol) and purified by column chromatography (1:9 ethyl acetate/hexane), yielding the product as a white solid (155.4 mg, 83%).

^{1}H-NMR	(400 MHz, CDCl$_3$, 25 °C): δ=8.23 (d, ^3J(H,H)=10.2 Hz, 1H, N-H), 7.60 (dd, ^3J(H,H)=14.6, 10.9 Hz, 1H H$_{arom}$), 7.29-7.32 (m, 2H H$_{arom}$), 7.25 (d, ^3J(H,H)=15.0 Hz, 2H H$_{arom}$), 7.16 (t, ^3J(H,H)=7.2 Hz, 1H, H-5), 6.26 (d, ^3J(H,H)=14.6 Hz, 1H, H-6), 5.84 (s, 1H, H-3), 5.46 (s, 1H, H-3), 2.03 (s, 3H, H-1) ppm.
^{13}C-NMR	(101 MHz, CDCl$_3$, 25 °C): δ=165.6 (C-4), 139.4 (C-2), 136.1 (C$_{arom}$), 128.6 (C$_{arom}$), 126.6 (C$_{arom}$), 125.5 (C$_{arom}$), 122.9 (C-5), 120.7 (C-3), 113.7 (C-6), 18.5 (C-1) ppm.
EA	Calcd: C, 76.98; H, 7.00; N, 7.48. Found: C, 75.27; H, 6.96; N, 7.21.
MS	(Ion trap, EI): *m/z* (%) = 187 (100, [*M*$^+$]), 159 (71), 144 (24), 117 (24), 69 (80), 41 (70).

Synthesis of 2-acetamido-N-[(E)-2-phenylvinyl]acetamide (10eb)

10eb was synthesized following method B using 2-acetylamido-acetamide (116.12 mg, 1.00 mmol) and phenylacetylene (220 µL, 2.00 mmol) and purified by column chromatography (1:5 ethyl acetate/hexane), yielding the product as a white solid (168 mg, 77%).

^{1}H-NMR	(600 MHz, CDCl$_3$, 25 °C): δ=8.47 (s, 1H, N-H), 7.42 (dd, ^3J(H,H)=14.5, 10.6 Hz, 1H, H-5), 7.25-7.30 (m, 4H, H$_{arom}$), 7.17 (t, ^3J(H,H)=7.0 Hz, 1H, H$_{arom}$), 6.44 (s, 1H, N-H), 6.20 (d, ^3J(H,H)=14.6 Hz, 1H, H-6), 4.03 (d, ^3J(H,H)=4.9 Hz, 2H, H-3), 2.08 (s, 3H, H-1) ppm.

Experimental part

^{13}C-NMR	(151 MHz, DMSO-d_6, 25 °C): δ=169.8 (C-4), 167.4 (C-2), 136.5 (C_{arom}), 128.7 (C_{arom}), 126.2 (C_{arom}), 125.2 (C_{arom}), 123.3 (C-5), 111.8 (C-6), 42.2 (C-3), 22.5 (C-1) ppm.
EA	Calcd: C, 66.04; H, 6.47; N, 12.84. Found: C, 65.79; H, 6.52; N, 12.54.
MS	(Ion trap, EI): *m/z* (%) = 218 (13, [M$^+$]), 119 (100), 91 (15), 77 (3), 65 (5), 43 (12).

Synthesis of ethyl oxo{[(*E*)-2-phenylvinyl]amino}acetate (10fb)

10fb was synthesized following method B using oxalamic acid ethyl ester (117.1 mg, 1.00 mmol) and phenylacetylene (220 µL, 2.00 mmol) and purified by column chromatography (1:7 ethyl acetate/hexane), yielding the product as a white solid (149.1 mg, 68%).

^{1}H-NMR	(400 MHz, CDCl$_3$, 25 °C): δ=9.00 (s, 1H, N-H), 7.47 (dd, ^3J(H,H)=14.6, 10.9 Hz, 1H, H-5), 7.35-7.39 (m, 2H, H_{arom}), 7.32 (t, ^3J(H,H)=7.7 Hz, 2H, H_{arom}), 7.24 (t, ^3J(H,H)=7.2 Hz, 1H, H_{arom}), 6.42 (d, ^3J(H,H)=14.6 Hz, 1H, H-6), 4.41 (q, ^3J(H,H)=7.2 Hz, 2H, H-2), 1.42 (t, ^3J(H,H)=7.2 Hz, 3H, H-1) ppm.
^{13}C-NMR	(101 MHz, CDCl$_3$, 25 °C): δ=160.2 (C-4), 153.4 (C-3), 135.1 (C_{arom}), 128.7 (C_{arom}), 127.4 (C_{arom}), 125.9 (C_{arom}), 121.0 (C-5), 116.9 (C-6), 63.5 (C-2), 13.9 (C-1) ppm.
EA	Calcd: C, 65.74; H, 5.98; N, 6.39. Found: C, 66.03; H, 5.95; N, 6.25.
MS	(Ion trap, EI): *m/z* (%) = 219 (100, [M$^+$]), 145 (39), 118 (26), 91 (13), 77 (5), 65 (4).

Synthesis of *N*-[(*E*)-2-(4-methoxyphenyl)vinyl]benzamide (Alatamide) (10ah)
(CAS-N0: 54797-23-8)

10ah was synthesized following method B benzamide (121.14 mg, 1.00 mmol) and 4-methoxy-phenylacetylene (259 µL, 2.00 mmol) and purified by column chromatography (1:7 ethyl acetate/hexane), yielding the product as a white solid (200 mg, 79%).

¹H-NMR (600 MHz, DMSO-d_6, 25 °C): δ=10.53 (d, ^3J(H,H)=9.7 Hz, 1H, N-H), 7.93-7.98 (m, 2H, H$_{arom}$), 7.55-7.61 (m, 1H, H-6), 7.46-7.54 (m, 3H, H$_{arom}$), 7.32 (d, ^3J(H,H)=8.5 Hz, 2H, H$_{arom}$), 6.85-6.92 (m, 2H, H$_{arom}$), 6.42 (d, ^3J(H,H)=15.0 Hz, 1H, H-7), 3.73 (s, 3H, H-12) ppm.

¹³C-NMR (151 MHz, DMSO-d_6, 25 °C): δ=163.8 (C-5), 158.0 (C$_{arom}$), 133.5 (C$_{arom}$), 131.8 (C$_{arom}$), 129.0 (C$_{arom}$), 128.4 (C$_{arom}$), 127.6 (C$_{arom}$), 126.4 (C$_{arom}$), 122.3 (C-6), 114.2 (C-7), 112.8 (C$_{arom}$), 55.1 (C-12) ppm.

MS (Ion trap, EI): m/z (%) = 253 (62, [M$^+$]), 150 (14), 133 (7), 105 (100), 77 (49), 51 (17).

Experimental part

5.4 Second generation catalyst for the addition of secondary amides to alkynes.

5.4.1 General method

An oven-dried flask was charged with potassium carbonate (20.7 mg, 0.15 mmol), and a stock solution containing ruthenium(III) chloride hydrate (7.8 mg, 0.03 mmol), 4-dimethylaminopyridine (11.0 mg, 0.09 mmol) in acetone (1.0 mL). The acetone was removed *in vacuo* and the flask was flushed with nitrogen. Subsequently, dry toluene (3.0 mL), water (7 µL, 0.40 mmol), *N*-nucleophile (1.00 mmol), tri-*n*-butylphosphine (22 µL, 0.09 mmol) and alkyne (2.00 mmol) were added *via* syringe. The resulting solution was stirred for 15 h at 100 °C, then poured into an aqueous sodium bicarbonate solution (30 mL). The resulting mixture was extracted repeatedly with 20 mL portions of ethyl acetate, the combined organic layers were washed with water and brine, dried with magnesium sulfate, filtered, and the volatiles were removed *in vacuo*. The residue was purified by column chromatography (silica gel, ethyl acetate/hexane) to yield the *E*-secondary enamide. The identity and purity of the products were confirmed by ^{1}H and ^{13}C NMR spectroscopy, mass spectroscopy and elemental analysis.

Experimental part

5.4.2 Synthesis of the enamides

Synthesis of N-((E)-hex-1-enyl)pyrrolidin-2-one (14aa) [CAS: 863709-29-9]

14aa was prepared from 1-hexyne (229 µL, 2.00 mmol) following the general procedure. The product was purified by column chromatography (silica gel, ethyl acetate/hexane 2:3) yielding a 32:1 mixture of **14a** and **15aa** as a colorless solid (165.6 mg, 99%).

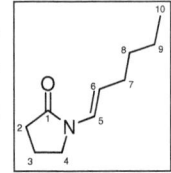

^{1}H-NMR (400 MHz, CDCl$_3$, 25 °C): δ=6.70 (d, ^3J(H,H)=14.5 Hz, 1H, H-5), 4.79 (dt, ^3J(H,H)=14.4 Hz, 7.2 Hz, 1H, H-6), 3.34 (t, ^3J(H,H)=7.2 Hz, 2H, H-4), 2.31 (t, ^3J(H,H)=8.2 Hz, 2H, H-2), 1.87-1.99 (m, 4H, H-7, 3), 1.12-1.27 (m, 4H, H-8, 9), 0.74 (t, ^3J(H,H)=7.2 Hz, 3H, H-10) ppm.

^{13}C-NMR (101 MHz, CDCl$_3$, 25 °C): δ=172.3 (C-1), 123.2 (C-5), 112.1 (C-6), 44.9 (C-4), 31.9 (C-7), 30.9 (C-2), 29.4 (C-8), 21.7 (C-9), 17.1 (C-3), 13.5 (C-10) ppm.

MS (Ion trap, EI): *m/z* (%) = 168 (12), 124 (100), 96 (44), 86 (23), 69 (11), 41 (14).

Synthesis of N-((E)-2-phenyl-vinyl)pyrrolidin-2-one (14ab) [CAS: 17234-64-9]

14ab was prepared from phenylacetylene (220 µL, 2.00 mmol) following the general procedure. The product was purified by column chromatography (silica gel, ethyl acetate/hexane 2:3) yielding a 22:1 mixture of **14ab** and **15ab** as a greyish solid (181.5.5 mg, 97%).

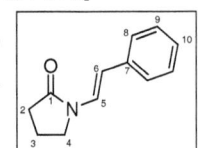

^{1}H-NMR (400 MHz, CDCl$_3$, 25 °C): δ=7.64 (d, ^3J(H,H)=14.6 Hz, 1H, H-5), 7.30-7.39 (m, 4H; H$_{arom}$), 7.19 (t, ^3J(H,H)=7.3 Hz, 1H, H$_{arom}$), 5.87 (d, ^3J(H,H)=15.0 Hz, 1H, H-6), 3.58 (t, ^3J(H,H)=7.3 Hz, 2H, H-4), 2.50 (t, ^3J(H,H)=8.2 Hz, 2H, H-2), 2.04-2.12 (m, 2H, H-3) ppm.

^{13}C-NMR (101 MHz, CDCl$_3$, 25 °C): δ=173.0 (C-1), 136.1 (C$_{arom}$), 128.3 (C$_{arom}$), 126.2 (C$_{arom}$), 125.3 (C$_{arom}$), 123.3 (C-5), 111.4 (C-6), 44.9 (C-4), 30.9 (C-2), 17.1 (C-3) ppm.

MS (Ion trap, EI): *m/z* (%) = 187 (100), 159 (11), 132 (54), 117 (17), 77 (12), 51 (7).

Experimental part

Synthesis of N-((E)-3,3-dimethyl-but-1-enyl)pyrrolidin-2-one (14ac) [CAS: 863709-44-8]

14ac was prepared from 3,3-dimethylbut-1-yne (246 µL, 2.00 mmol) following the general procedure. The product was purified by column chromatography (silica gel, ethyl acetate/hexane 2:3) yielding a 40:1 mixture of **14ac** and **15ac** as a colorless solid (167.0 mg, 99%).

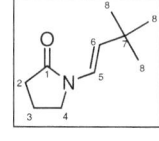

¹H-NMR (400 MHz, CDCl₃, 25 °C): δ=6.68 (d, ³J(H,H)=15.0 Hz, 1H, H-5), 4.85 (d, ³J(H,H)=14.7 Hz, 1H; H-6), 3.34 (t, ³J(H,H)=7.2 Hz, 2H, H-4), 2.31 (t, ³J(H,H)=8.2 Hz, 2H, H-2), 1.89-1.98 (m, 2H, H-3), 0.91 (s, 9H, H-8) ppm.

¹³C-NMR (101 MHz, CDCl₃, 25 °C): δ=172.5 (C-1), 123.5 (C-5), 120.1 (C-6), 45.0 (C-4), 31.7 (C-7), 31.1 (C-2), 30.0 (C-3), 17.1 (C-8) ppm.

MS (Ion trap, EI): m/z (%) = 167 (9), 152 (100), 134 (18), 67 (11), 41 (9).

Synthesis of N-((E)-4-phenyl-but-1-enyl)pyrrolidin-2-one (14ad) [CAS: 863709-36-8]

14ad was prepared from (but-3-ynyl)benzene (281 µL, 2.00 mmol) following the general procedure. The product was purified by column chromatography (silica gel, ethyl acetate/hexane 1:4) yielding a 25:1 mixture of **14ad** and **15ad** as a colorless oil (206.5 mg, 96%).

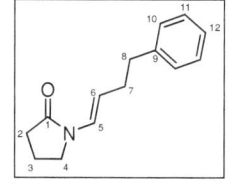

¹H-NMR (400 MHz, CDCl₃, 25 °C): δ=7.18-7.25 (m, 2H, H_arom), 7.09-7.15 (m, 3H, H_arom), 6.87 (d, ³J(H,H)=14.4 Hz, 1H, H-5), 4.89 (dt, ³J(H,H)=14.4 Hz, 7.2 Hz, 1H, H-6), 3.35 (t, ³J(H,H)=7.2 Hz, 2H, H-4), 2.60-2.68 (m, 2H, H-8), 2.30-2.39 (m, 4H, H-2, 7), 1.90-1.99 (m, 2H, H-3) ppm.

¹³C-NMR (101 MHz, CDCl₃, 25 °C): δ=172.2 (C-1), 141.6 (C-9), 128.0 (C_arom), 127.9 (C_arom), 125.5 (C_arom), 123.8 (C-5), 110.7 (C-6), 44.8 (C-4), 36.3 (C8), 31.6 (C-7), 30.8 (C-2), 17.0 (C-3).

MS (Ion trap, EI): m/z (%) = 215 (1), 124 (100), 96 (28), 69 (7), 41 (8).

Experimental part

Synthesis of methyl-(E)-3-(2-oxopyrrolidin-1-yl) acrylate (14ae) [CAS: 145294-78-6]

14ae was prepared from methyl propargylate (179 µL, 2.00 mmol) following the general procedure. The product was purified by column chromatography (silica gel, ethyl acetate/hexane 1:4) yielding a 40:1 mixture of **14ae** and **15ae** as a colorless solid (162.3 mg, 96%).

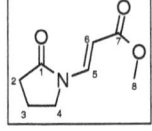

¹H-NMR (400 MHz, CDCl$_3$, 25 °C): δ=7.94 (d, ^3J(H,H)=14.3 Hz, 1H, H-5), 5.07 (d, ^3J(H,H)=14.1 Hz, 1H; H-6), 3.59 (s, 3H, H-8), 3.44 (t, ^3J(H,H)=7.2 Hz, 2H, H-4), 2.382.45 (m, 2H, H-2), 2.02-2.09 (m, 2H, H-3) ppm.

¹³C-NMR (101 MHz, CDCl$_3$, 25 °C): δ=174.0 (C-1), 167.3 (C-7), 137.1 (C-5), 99.7 (C-6), 51.0 (C-8), 44.6 (C-4), 30.6 (C-2), 17.1 (C-3) ppm.

MS (Ion trap, EI): *m/z* (%) = 169 (43), 138 (77), 110 (100), 82 (74), 70 (27), 55 (14), 41 (11).

Synthesis of N-((E)-3-methyl-1,3-butadien-1-yl)pyrrolidin-2-one (14af) [CAS: 112682-83-4]

14af was prepared from 2-methylbut-1-en-3-yne (192 µL, 2.00 mmol) following the general procedure. The product was purified by column chromatography (silica gel, ethyl acetate/hexane 1:4) yielding a 8:1 mixture of **14af** and **15af** as a colorless solid (148.8 mg, 97%).

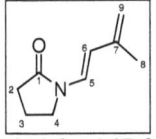

¹H-NMR (400 MHz, CDCl$_3$, 25 °C): δ=6.94 (d, ^3J(H,H)=14.7 Hz, 1H, H-5), 5.61 (d, ^3J(H,H)=14.7 Hz, 1H; H-6), 4.79(s, 1H, H-9), 4.75 (s, 1H, H-9), 3.45-3.47 (m, 2H, H-4), 2.40 (t, ^3J(H,H)=8.2 Hz, 2H, H-2), 1.99-2.04 (m, 2H, H-3), 1.77 (s, 3H, H-8) ppm.

¹³C-NMR (101 MHz, CDCl$_3$, 25 °C): δ=173.1 (C-1), 140.3 (C-7), 123.3 (C-5), 114.9 (C-9), 114.2 (C-6), 45.0 (C-4), 31.0 (C-2), 18.5 (C-3), 17.1 (C-8) ppm.

MS (Ion trap, EI): *m/z* (%) = 153 (27), 138 (41), 110 (100), 82 (45), 69 (15), 43 (20).

Experimental part

Synthesis of *N*-((*E*)-5-chloro-pent-1-enyl)pyrrolidin-2-one (3ag) [CAS: 863709-31-3]

14ag was prepared from 5-chloropent-1-yne (212 µL, 2.00 mmol) following the general procedure. The product was purified by column chromatography (silica gel, ethyl acetate/hexane 2:3) yielding a 33:1 mixture of **14ag** and **15ag** as a colorless oil (186.0 mg, 99%).

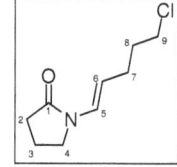

¹H-NMR (400 MHz, CDCl₃, 25 °C): δ=6.76 (d, ³*J*(H,H)=14.5 Hz, 1H, H-5), 4.76 (dt, ³*J*(H,H)=14.5 Hz, 7.2 Hz, 1H, H-6), 3.34-3.39 (m, 4H, H-4, 9), 2.32 (t, ³*J*(H,H)=8.1 Hz, 2H, H-2), 2.06-2.11 (m, 2H, H-7), 1.92-1.99 (m, 2H, H-3), 1.68-1.74(m, 2H, H-8) ppm.

¹³C-NMR (101 MHz, CDCl₃, 25 °C): δ=172.5 (C-1), 124.3 (C-5), 109.6 (C-6), 44.9 (C-4), 43.8 (C-9), 32.4 (C-7), 30.9 (C-2), 26.9 (C-8), 17.0 (C-3) ppm.

MS (Ion trap, EI): *m/z* (%) = 187 (18), 124 (100), 96 (34), 69 (10), 41 (12).

Synthesis of *N*-((*E*)-3-methoxy-prop-1-enyl)pyrrolidin-2-one (14ah) [CAS: 863709-43-7]

14ah was prepared from 3-methoxypropyne (169 µL, 2.00 mmol) following the general procedure. The product was purified by column chromatography (silica gel, ethyl acetate/hexane 1:4) yielding a 24:1 mixture of **14ah** and **15ah** as a colorless oil (154.5 mg, 99%).

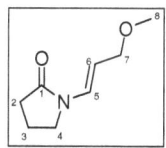

¹H-NMR (400 MHz, CDCl₃, 25 °C): δ=6.93 (d, ³*J*(H,H)=14.5Hz, 1H, H-5), 4.89 (dt, ³*J*(H,H)=14.4 Hz, 7.1 Hz, 1H; H-6), 3.77-3.81 (m, 2H, H-7), 3.36-3.40 (m, 2H, H-4), 3.15 (s, 3H, H-8), 2.33 (t, ³*J*(H,H)=8.1 Hz, 2H, H-2), 1.93-2.00 (m, 2H, H-3) ppm.

¹³C-NMR (101 MHz, CDCl₃, 25 °C): δ=173.0 (C-1), 126.9 (C-5), 106.8 (C-6), 70.9 (C-7), 57.0 (C-8), 44.7 (C-4), 30.8 (C-2), 17.0 (C-3) ppm.

MS (Ion trap, EI): *m/z* (%) = 155 (33), 140 (22), 124 (100), 112 (54), 96 (37), 69 (26), 56 (9), 41 (28).

Experimental part

Synthesis of N-((E)-2-trimethylsilyl-vinyl)pyrrolidin-2-on (14ai) [CAS: 713147-64-9]

14ai was prepared from trimethylsilylacetylene (2i, 277 µL, 2.00 mmol) following the general procedure. The product was purified by column chromatography (silica gel, ethyl acetate/hexane 2:3) yielding a 2:1 mixture of 14ai and 15ai as a colorless oil (164.8 mg, 90%).

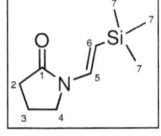

^{1}H-NMR (400 MHz, CDCl$_3$, 25 °C): δ=7.01 (d, ^3J(H,H)=17.3 Hz, 1H, H-5), 4.67(d, ^3J(H,H)=17.1 Hz, 1H; H-6), 3.47 (t, ^3J(H,H)=7.2 Hz, 2H, H-4), 2.40 (t, ^3J(H,H)=8.2 Hz, 2H, H-2), 1.95-2.04 (m, 2H, H-3), 0.90 (s, 9H, H-7) ppm.

^{13}C-NMR (101 MHz, CDCl$_3$, 25 °C): δ=172.8 (C-1), 133.4 (C-5), 105.9 (C-6), 44.4 (C-4), 31.5 (C-2), 17.1 (C-3), -1.0 (C-7) ppm.

MS (Ion trap, EI): m/z (%) = 183 (4), 168 (100), 142 (44), 100 (5), 75 (5), 59 (5).

Synthesis of N-((E)-dodeca-1,11-dienyl)pyrrolidin-2-one (14aj)

14aj was prepared from dodec-1-en-11-yne (347 µL, 1.5 mmol) following the general procedure. The product was purified by column chromatography (silica gel, ethyl acetate/hexane 2:3) yielding a 29:1 mixture of 14aj and 15aj as a yellowish oil (178.3 mg, 71.5%).

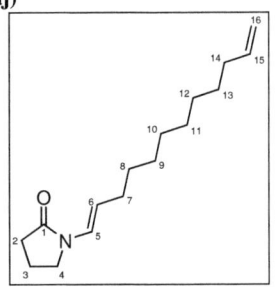

^{1}H-NMR (400 MHz, CDCl$_3$, 25 °C): δ=6.75 (d, ^3J(H,H)=14.5 Hz, 1H, H-5), 5.68 (qd, ^3J(H,H)=17.0 Hz, 10.2 Hz, 6.71 Hz, 1H, H-15), 5.13-5.45 (m, 1H, H-6), 4.77-4.90 (m, 2H, H-16), 3.38 (t, ^3J(H,H)=7.1 Hz, 2H, H-4), 2.32-2.38 (m, 2H, H-2), 1.88-2.03 (m, 6H, H-3, 7, 14), 1.10-1.31 (m, 12H, H-8, 9, 10, 11, 12, 13) ppm.

^{13}C-NMR (101 MHz, CDCl$_3$, 25 °C): δ=172.4 (C-1), 138.8 (C-15), 123.3 (C-5), 113.8 (C-6), 112.1 (C-16), 45.0 (C-4), 33.5, 31.0, 29.8, 29.3, 29.1, 28.8, 28.6 (C-2, 7, 8, 9, 10, 11, 12, 13, 14), 17.1 (C-3) ppm.

MS (Ion trap, EI): m/z (%) = 249 (2), 166 (7), 124 (100), 96 (33), 86 (62), 69 (13), 41 (18).

Experimental part

Synthesis of *N*-((*E*)-hex-1-enyl)azetidin-2-one (14ba) [CAS: 863709-38-0]

14ba was prepared from azetidin-2-one (71.1 mg, 1.00 mmol) following the general procedure. The product was purified by column chromatography (silica gel, ethyl acetate/hexane 2:3) yielding a 1:1 mixture of **14ba** and **15ab** as a yellowish oil (53.6 mg, 35%).

¹H-NMR   (400 MHz, CDCl₃, 25 °C): δ=6.44 (d, ³*J*(H,H)=14.3 Hz, 1H, H-4), 4.96 (dt, ³*J*(H,H)=14.3 Hz, 7.1 Hz, 1H, H-5), 3.32 (t, ³*J*(H,H)=4.4 Hz, 2H, H-3), 2.91 (t, ³*J*(H,H)=4.5 Hz, 2H, H-2), 1.96 (q, ³*J*(H,H)=6.7 Hz, 2H, H-6), 1.15-1.37 (m, 4H, H7, 8), 0.80-0.89 (m, 3H, H9) ppm.

¹³C-NMR   (101 MHz, CDCl₃, 25 °C): δ=163.5 (C-1), 121.5 (C-4), 111.8 (C-5), 38.0 (C-3), 35.7 (C-2), 31.8 (C-6), 29.0 (C-7), 21.9 (C-8), 13.8 (C-9) ppm.

MS   (Ion trap, EI): *m/z* (%) = 153 (10), 124 (5), 110 (60), 82 (29), 68 (100), 55 (11), 41 (31).

Synthesis of *N*-((*E*)-hex-1-enyl)piperidin-2-one (14ca) [CAS: 863709-45-9]

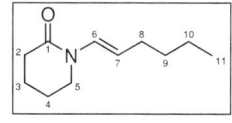

14ca was prepared from piperidin-2-one (99.1 mg, 1.00 mmol) following the general procedure. The product was purified by column chromatography (silica gel, ethyl acetate/hexane 2:3) yielding a 40:1 mixture of **14ca** and **15ca** as a yellowish oil (165.1 mg, 91%).

¹H-NMR   (400 MHz, CDCl₃, 25 °C): δ=7.27 (d, ³*J*(H,H)=14.7 Hz, 1H, H-6), 4.91 (dt, ³*J*(H,H)=14.5 Hz, 7.1 Hz, 1H, H-7), 3.28 (t, ³*J*(H,H)=6.1 Hz, 2H, H-5), 2.36 (t, ³*J*(H,H)=6.6 Hz 2H, H-2), 1.91-2.09 (m, 2H, H-8), 1.62-1.86 (m, 4H, H-3, 4), 1.17-1.34 (m, 4H, H-9, 10), 0.79 (t, ³*J*(H,H)=7.2 Hz, 3H, H-11) ppm.

¹³C-NMR   (101 MHz, CDCl₃, 25 °C): δ=167.7 (C-1), 126.4 (C6), 111.1 (C-7), 44.8 (C-5), 32.5 (C-2), 32.1 (C-8), 29.7 (C-9), 22.3 (C-10), 21.8 (C-4), 20.3 (C-3), 13.6 (C-11) ppm.

MS   (Ion trap, EI): *m/z* (%) = 181 (16), 152 (20), 138 (100), 124 (58), 110 (39), 100 (32), 82 (24), 55 (19), 41 (14).

Experimental part

Synthesis of N-((E)-hex-1-enyl)azepan-2-one (14da)

14da was prepared from azepan-2-one (113.2 mg, 1.00 mmol) following the general procedure. The product was purified by column chromatography (silica gel, ethyl acetate/hexane 2:3) yielding a 40:1 mixture of **14da** and **15da** as a yellowish oil (194.6 mg, 99%).

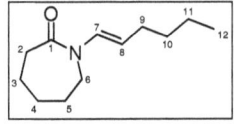

^{1}H-NMR (400 MHz, CDCl$_3$, 25 °C): δ=6.96 (d, ^3J(H,H)=14.5 Hz, 1H, H-7), 4.87 (dt, ^3J(H,H)=14.5 Hz, 7.1 Hz, 1H, H-8), 3.41-3.45 (m, 2H, H-6), 2.42-2.49 (m, 2H, H-2), 1.89-1.97 (m, 2H, H-9), 1.47-1.64 (m, 6H, H-3, 4, 5), 1.13-1.26 (m, 4H, H-10, 11), 0.74 (t, ^3J(H,H)=7.2 Hz, 3H, H-12) ppm.

^{13}C-NMR (101 MHz, CDCl$_3$, 25 °C): δ=173.4 (C-1), 126.2 (C-7), 110.7 (C-8), 45.1 (C-6), 36.8 (C-2), 32.1 (C-9), 29.6 (C-10), 29.1 (C-5), 27.0 (C-4), 23.1 (C-3), 21.8 (C-11), 13.6 (C-12) ppm.

EA Calcd: C, 73.8; H, 10.8; N, 7.2.
Found: C, 73.9; H, 11.0; N, 7.2.

MS (Ion trap, EI): m/z (%) = 196 (42), 166 (12), 152 (63), 138 (100), 124 (22), 114 (16), 110 (10), 96 (20), 84 (11), 69 (10), 41 (18).

Synthesis of N-((E)-hex-1-enyl)azonan-2-one (14ea) [CAS: 863709-32-4]

14ea was prepared from azonan-2-one (141.2 mg, 1.00 mmol) following the general procedure. The product was purified by column chromatography (silica gel, ethyl acetate/hexane 2:3) yielding a 40:1 mixture of **14ea** and **15ea** as a yellowish oil (67.1 mg, 30%).

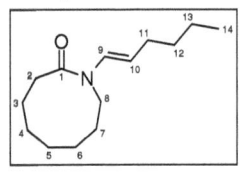

^{1}H-NMR (400 MHz, CDCl$_3$, 25 °C): δ=7.16 (d, ^3J(H,H)=14.7 Hz, 1H, H-9), 4.97 (dt, ^3J(H,H)=14.6 Hz, 7.2 Hz, 1H, H-10), 3.68 (t, ^3J(H,H)=5.7 Hz, 2H, H-8), 2.52-2.56 (m, 2H, H-2), 2.00-2.06 (m, 2H, H-11), 1.75-1.82 (m, 2H, H-3), 1.62-1.70 (m, 2H, H-7), 1.37-1.55 (m, 6H, H-4, 5, 6), 1.22-1.35 (m, 4H, H-12, 13), 0.84 (t, ^3J(H,H)=7.2 Hz, 3H, H-14) ppm.

Experimental part

¹³C-NMR (101 MHz, CDCl₃, 25 °C): δ=173.5 (C-1), 125.0 (C-9), 112.3 (C-10), 45.4 (C-8), 35.2 (C-2), 32.3 (C-7), 30.1 (C-11), 28.2 (C-12), 25.6 (C-6), 25.2 (C-5), 25.1 (C-3), 22.1 (C-4), 22.0 (C-13), 13.8 (C-14) ppm.

MS (Ion trap, EI): *m/z* (%) = 223 (32), 180 (100), 166 (100), 112 (32), 55 (29), 41 (27).

Synthesis of 1,4-di-((*E*)-hex-1-enyl)piperazin-2,5-dione (14fa) [CAS: 863709-41-5]

14fa was prepared from piperazin-2,5-dione (114.1 mg, 1.00 mmol) and 1-hexyne (2a, 456 µL, 4.00 mmol) following the general procedure. The product was purified by column chromatography (silica gel, ethyl acetate/hexane 2:3) yielding a 40:1 mixture of 14fa and 15fa as a colorless solid (196.4 mg, 71%).

¹H-NMR (400 MHz, CDCl₃, 25 °C): δ=7.18 (d, ³*J*(H,H)=14.5 Hz, 2H, H-3), 5.10 (dt, ³*J*(H,H)=14.6 Hz, 7.2 Hz, 2H, H-4), 4.14 (s, 4H, H-2), 2.09 (q, ³*J*(H,H)=6.9 Hz, 4H, H-5), 1.26-1.40 (m, 8H, H-6, 7), 0.88 (t, ³*J*(H,H)=7.2 Hz, 6H, H-8) ppm.

¹³C-NMR (101 MHz, CDCl₃, 25 °C): δ=160.1 (C-1), 123.6 (C-3), 114.8 (C-4), 47.4 (C-2), 31.9 (C-5), 29.7 (C-6), 22.0 (C-7), 13.8 (C-8) ppm.

MS (Ion trap, EI): *m/z* (%) = 278 (56), 235 (58), 207 (100), 112 (29), 68 (42), 41 (32).

Synthesis of *N*-(4-acetyl-phenyl)-*N*-((*E*)-hex-1-enyl)acetamide (14ga)
[CAS: 863709-40-49]

14ga was prepared from *N*-(4-acetyl-phenyl)acetamide (177.2 mg, 1.00 mmol) following the general procedure. The product was purified by column chromatography (silica gel, ethyl acetate/hexane 2:3) yielding a 40:1 mixture of 14ga and 15ga as a colorless solid (50.5 mg, 20%).

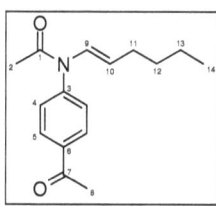

¹H-NMR (400 MHz, CDCl₃, 25 °C): δ=8.07 (d, ³*J*(H,H)=8.4 Hz, 2H, H$_{arom}$), 7.41 (d, ³*J*(H,H)=8.2 Hz, 2H, H$_{arom}$), 7.27 (d, ³*J*(H,H)=13.1 Hz, 1H, H-9), 4.29 (dt,

Experimental part

^3J(H,H)=13.9 Hz, 6.8 Hz, 1H, H-10), 2.61 (s, 3H, H-8), 1.91-1.97 (m, 2H, H-11), 1.81 (s, 3H, H-2), 1.13-1.23 (m, 4H, H-12,13), 0.78 (m, 3H, H-14) ppm.

^{13}C-NMR (101 MHz, CDCl$_3$, 25 °C): 197.2 (C-7), 175.1 (C-1), 143.9 (C$_{arom}$), 136.3 (C$_{arom}$), 129.9 (C$_{arom}$), 129.2 (C$_{arom}$), 128.3 (C-9), 113.8 (C-10), 31.4 (C-11), 29.6 (C-12), 26.8 (C-8), 23.0 (C-2), 21.4 (C-13), 13.7 (C-14) ppm.

MS (Ion trap, EI): m/z (%) = 259 (16), 217 (14), 202 (19), 174 (100), 132 (27), 77 (7), 43 (29).

Synthesis of *N*-(4-ethoxy-phenyl)-*N*-((*E*)-hex-1-enyl)acetamide (14ha)
[CAS: 863709-47-1]

14ha was prepared from *N*-(4-ethoxy-phenyl)acetamide (179.2 mg, 1.00 mmol) following the general procedure. The product was purified by column chromatography (silica gel, ethyl acetate/hexane 2:3) yielding a 40:1 mixture of **14ha** and **15ha** as a colorless solid (234.2 mg, 93%).

^{1}H-NMR (400 MHz, CDCl$_3$, 25 °C): δ=7.33 (d, ^3J(H,H)=14.5 Hz, 1H, H-1), 6.93 (d, ^3J(H,H)=8.8 Hz, 2H, H$_{arom}$), 6.84 (d, ^3J(H,H)=8.8 Hz, 2H, H$_{arom}$), 4.29 (dt, ^3J(H,H)=14.3 Hz, 7,2 Hz, 1H, H-10), 3.94 (q, ^3J(H,H)=6.9 Hz, 2H, H-7), 1.85 (q, ^3J(H,H)=6.8 Hz, 2H, H-11), 1.71 (s, 3H, H-2), 1.32 (t, ^3J(H,H)=6.9 Hz, 3H, H-8), 1.05-1.19 (m, 4H, H-12, 13), 0.71 (t, ^3J(H,H)=7.0 Hz, 3H, H-14) ppm.

^{13}C-NMR (101 MHz, CDCl$_3$, 25 °C): 168.3 (C-1), 158.4 (C$_{arom}$), 132.3 (C$_{arom}$), 129.4 (C$_{arom}$), 128.1 (C-9), 115.2 (C$_{arom}$), 114.0 (C-10), 63.4 (C-7), 31.8 (C-11), 29.3 (C-12), 22.8 (C-2), 21.8 (C-13), 14.5 (C-8), 13.6 (C-14) ppm.

MS (Ion trap, EI): m/z (%) = 261 (32), 218 (20), 204 (16), 176 (100), 148 (21), 108 (5), 65 (5), 43 (12).

Experimental part

Synthesis of N-phenyl-N-((E)-hex-1-enyl)benzamide (14ia)

14ia was prepared from benzanilide (197.2 mg, 1.00 mmol) following the general procedure. The product was purified by column chromatography (silica gel, ethyl acetate/hexane 2:3) yielding a 40:1 mixture of **14ia** and **15ia** as a colorless solid (56.1 mg, 20%).

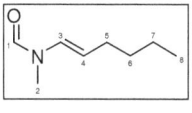

¹H-NMR (400 MHz, CDCl₃, 25 °C): δ=7.36 (d, ³J(H,H)=10.0 Hz, 1H, H-10), 7.25-7.34 (m, 4H, H_arom), 7.12-7.24 (m, 4H, H_arom), 7.09 (d, ³J = 7.4 Hz, 2H, H_arom), 4.72 (dt, ³J(H,H)=14.3 Hz, 7,2 Hz, 1H, H-11), 2.03 (q, ³J(H,H)=7.0 Hz, 2H, H-12), 1.19-1.35 (m, 4H, H-13, 14), 0.85 (t, ³J(H,H)=7.1 Hz, 3H, H-15) ppm.

¹³C-NMR (101 MHz, CDCl₃, 25 °C): 168.6 (C-1), 140.3 (C_arom), 135.8 (C_arom), 129.6 (C_arom), 129.4 (C_arom), 129.2 (C_arom), 129.1 (C_arom), 128.5 (C-10), 127,7 (C_arom), 127.5 (C_arom), 116.6 (C-11), 31.9 (C-12), 29.7 (C-13), 22.1 (C-14), 13.9 (C-15) ppm.

EA Calcd: C, 81.7; H, 7.6; N, 5.0.
Found: C, 81.7; H, 7.5; N, 4.9.

MS (Ion trap, EI): *m/z* (%) = 279 (11), 222 (16), 197 (7), 174 (10), 105 (100), 77 (39), 51 (11).

Synthesis of N-((E)-hex-1-enyl)-N-(methyl)formamide (14ja) [CAS: 863709-32-4]

14ja was prepared from N-methylformamide (59.1 mg, 1.00 mmol) following the general procedure. The product was purified by column chromatography (silica gel, ethyl acetate/hexane 2:3) yielding a 3:1 mixture of **14ja** and **15ja** as a yellowish oil (48.7 mg, 35%).

¹H-NMR (400 MHz, CDCl₃, 25 °C): δ=8.23 (s, 1H, H-1), 6.41 (d, ³J(H,H)=14.0 Hz, 1H, H-3), 5.01-5.11 (m, 1H, H-4), 2.97 (s, 3H, H-2), 2.00-2.10 (m, 2H, H-5), 1.20-1.41 (m, 4H, H-6, 7), 0.84-0.91 (m, 3H, H-8) ppm.

¹³C-NMR (101 MHz, CDCl₃, 25 °C): 162.0 (C-1), 128.1 (C-3), 111.5 (C-4), 32.2 (C-2), 29.7 (C-5), 27.4 (C-6), 22.0 (C-7), 13.8 (C-8) ppm.

MS (Ion trap, EI): *m/z* (%) = 141 (5), 98 (100), 84 (11), 70 (83), 60 (23), 42 (35).

Experimental part

Synthesis of N-((E)-hex-1-enyl)-N-(methyl)acetamide (14ka) [CAS: 863709-46-0]

14ka was prepared from N-methylacetamide (73.1 mg, 1.00 mmol) following the general procedure. The product was purified by column chromatography (silica gel, ethyl acetate/hexane 2:3) yielding a 7:1 mixture of **14ka** and **15ka** as a yellowish oil (14.7 mg, 10%).

¹H-NMR (400 MHz, CDCl$_3$, 25 °C): δ=6.57 (d, ^3J(H,H)=13.8 Hz, 1H, H-4), 4.94-5.01 (m, 1H, H-5), 3.04 (s, 3H, H-3), 2.17 (s, 3H, H-2), 2.00-2.08 (m, 2H, H-5), 1.26-1.39 (m, 4H, H-7, 8), 0.89 (t, ^3J(H,H)=7.1 Hz, 3H, H-9) ppm.

¹³C-NMR (101 MHz, CDCl$_3$, 25 °C): 168.9 (C-1), 128.6 (C-4), 112.3 (C-5), 32.4 (C-3), 30.1 (C-6), 29.6 (C-7), 22.1 (C-8), 22.0 (C-2), 13.9 (C-9) ppm.

MS (Ion trap, EI): *m/z* (%) = 155 (8), 112 (22), 98 (20), 70 (100), 43 (18).

Synthesis of N-((E)-hex-1-enyl)-N-(isopropyl)acrylamide (14la) [CAS: 863709-34-6]

14la was prepared from N-isopropylacrylamide (113.2 mg, 1.00 mmol) following the general procedure. The product was purified by column chromatography (silica gel, ethyl acetate/hexane 2:3) yielding a 40:1 mixture of **14la** and **15la** as a colorless oil (96.1 mg, 49%).

¹H-NMR (400 MHz, CDCl$_3$, 25 °C): δ=6.54 (dd, $^3J_{trans}$(H,H)=16.9 Hz, $^3J_{cis}$(H,H)=10.3 Hz, 1H, H-9), 6.27 (dd, $^3J_{trans}$(H,H)=16.9 Hz, $^2J_{gem}$(H,H)=1.4 Hz, 1H, H-3), 5.92 (d, ^3J(H,H)=13.6 Hz, 1H, H-6), 5.54 (dd, $^3J_{cis}$(H,H)=10.3 Hz, $^2J_{gem}$(H,H)=1.2 Hz 1H, H-9), 5.44 (dt, ^3J(H,H)=13.8 Hz, 7.0 Hz, 1H, H-7), 4.70-4.83 (m, 1H, H-4), 2.11 (q, ^3J(H,H)=7.2 Hz, 2H, H-8), 1.28-1.43 (m, 4H, H-9, 10), 1.10 (d, ^3J(H,H)=6.8 Hz, 6H, H-5), 0.90 (t, ^3J(H,H)=7.1 Hz, 3H, H-11) ppm.

¹³C-NMR (101 MHz, CDCl$_3$, 25 °C): 165.2 (C-1), 134.7 (C-2), 130.0 (C-3), 126.3 (C-6), 123.8 (C-7), 45.4 (C-4), 31.3 (C-8), 29.7 (C-9), 22.2 (C-10), 20.0 (C-5), 13.8 (C-11) ppm.

MS (Ion trap, EI): *m/z* (%) = 196 (61), 180 (18), 152 (45), 138 (97), 110 (44), 98 (100), 55 (40).

Experimental part

Synthesis of 3-((*E*)-hex-1-enyl)oxazolidin-2-one (14ma) [CAS: 888973-38-4]

14ma was prepared from oxazolidin-2-one (87.1 mg, 1.00 mmol) following the general procedure. The product was purified by column chromatography (silica gel, ethyl acetate/hexane 2:3) yielding a 30:1 mixture of **14ma** and **15ma** as a colorless oil (148.6 mg, 88%).

^{1}H-NMR (400 MHz, CDCl$_3$, 25 °C): δ=6.50 (d, ^3J(H,H)=14.3 Hz, 1H; H-4), 4.72 (dt, ^3J(H,H)=14.3 Hz, 7.1 Hz, 1H, H-5), 4.32 (t, ^3J(H,H)=8.1 Hz, 2H, H-2), 3.59 (t, ^3J(H,H)=8.1 Hz, 2H, H-3), 1.96 (dt, ^3J(H,H)=14.4 Hz, 7.1 Hz, 2H, H-6), 1.18-1.30 (m, 4H, H-7, 8), 0.79 (t, ^3J(H,H)=7.1 Hz, 3H, H-9) ppm.

^{13}C-NMR (101 MHz, CDCl$_3$, 25 °C): 155.2 (C-1), 123.4 (C-4), 111.1 (C-5), 61.9 (C-2), 42.4 (C-3), 32.1 (C-6), 29.2 (C-7), 21.7 (C-8), 13.6 (C-9) ppm.

MS (Ion trap, EI): *m/z* (%) = 170 (19), 126 (100), 88 (14), 55 (5), 42 (46).

Synthesis of (4*R*)-3-((*E*)-hex-1-enyl)-4-(isopropyl)-oxazolidin-2-one (14na)

14na was prepared from 4(*R*)-isopropyl-oxazolidin-2-one (129.2 mg, 1.00 mmol) following the general procedure. The product was purified by column chromatography (silica gel, ethyl acetate/hexane 2:3) yielding a 22:1 mixture of **14ma** and **15ma** as a colorless oil (198.8 mg, 94.1%).

^{1}H-NMR (400 MHz, CDCl$_3$, 25 °C): δ=6.36 (d, ^3J(H,H)=14.5 Hz, 1H; H-6), 4.86 (dt, ^3J(H,H)=14.5 Hz, 7.2 Hz, 1H, H-7), 4.04-4.14 (m, 2H, H-2), 3.91 (dt, ^3J(H,H)=8.6 Hz, 3.4 Hz 1H, H-3), 2.21-2.33 (m, 1H, H-4), 1.88-1.99 (m, 2H, H-8), 1.15-1.29 (m, 4H, H-9, 10), 0.74-0.85 (m, 6H, H-5, 11), 0.70 (d, ^3J(H,H)=6.8 Hz, 3H, H-5) ppm.

^{13}C-NMR (101 MHz, CDCl$_3$, 25 °C): 155.3 (C-1), 122.2 (C-6), 111.9 (C-7), 62.6 (C-2), 58.1 (C-3), 31.8 (C-8), 29.5 (C-9), 25.7 (C-4), 21.7 (C-10), 17.5 (C-5), 13.6 (C-11), 13.5 (C-5) ppm.

MS (Ion trap, EI): *m/z* (%) = 212 (100), 168 (47), 124 (16), 56 (11), 41 (5).

Experimental part

Synthesis of (4S,5R)-3-((E)-hex-1-enyl)-5-methyl-4-phenyloxazolidin-2-one (14oa)
[CAS: 863709-49-3]

14oa was prepared from (4S,5R)-5-methyl-4-phenyloxazolidin-2-one (177.2 mg, 1.00 mmol) following the general procedure. The product was purified by column chromatography (silica gel, ethyl acetate/hexane 2:3) yielding a 21:1 mixture of 14oa and 15oa as a colorless solid (243.3 mg, 94%).

¹H-NMR	(400 MHz, CDCl₃, 25 °C): δ=7.18-7.35 (m, 5H, H$_{arom}$), 6.46 (d, ³J(H,H)=14.7 Hz, 1H; H-9), 5.55 (d, ³J(H,H)=7.7 Hz, 1H, H-3), 4.87 (dt, ³J(H,H)=14.5 Hz, 7.2 Hz, 1H, H-10), 4.27 (dt, ³J(H,H)=13.8 Hz, 6.7 Hz, 1H, H-2), 1.97 (q, ³J(H,H)=6.8 Hz, 2H, H-11), 1.18-1.35 (m, 4H, H-12, 13), 0.81 (t, ³J(H,H)=7.2 Hz, 3H, H-14), 0.72 (d, ³J(H,H)=6.4 Hz, 3H, H-4) ppm.
¹³C-NMR	(101 MHz, CDCl₃, 25 °C): 154.1 (C-1), 134.0 (C$_{arom}$), 128.2 (C$_{arom}$), 128.1 (C-8), 125.5 (C$_{arom}$), 122.1 (C-9), 112.0 (C-10), 78.6 (C-2), 53.9 (C-3), 31.8 (C-11), 29.4 (C-12), 21.7 (C-13), 13.6 (C-4), 12.7 (C-14) ppm.
MS	(Ion trap, EI): m/z (%) = 259 (12), 216 (24), 172 (100), 118 (78), 91 (45), 55 (24), 41 (17).

Synthesis of N-[(E)-hex-1-enyl]succinimide (14pa (4aa))

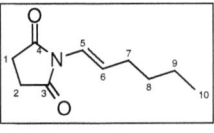

14pa was prepared from succinimide (99.1 mg, 1.00 mmol) following the general procedure. The product was purified by column chromatography (silica gel, ethyl acetate/hexane 2:3) yielding a 21:1 mixture of 14pa and 15pa as a colorless solid (243.3 mg, 94%).

¹H-NMR	(400 MHz, CDCl₃, 25 °C): δ=6.52 (d, ³J(H,H)=14.7 Hz, 1H, H-5), 6.36 (dt, ³J(H,H)=15.0 Hz, 14.7 Hz, 1H, H-6), 2.66 (s, 4H, H-1, H-2), 2.05 (dt, ³J(H,H)=14.6, 7.2, 2H, H-7), 1.24 – 1.37 (m, 4H, H-8, H-9), 0.83 (t, ³J(H,H)=7.2 Hz, 3H, H-10) ppm.
¹³C-NMR	(101 MHz, CDCl₃, 25 °C): δ=175.4 (C-3, C-4), 124.5 (C-5), 117.8 (C-6), 31.2 (C-7), 30.5 (C-8), 27.6 (C-1, C-2), 22.0 (C-9), 13.7 (C-10) ppm.
MS	(Ion trap, EI): m/z (%) = 182 (30, [M⁺]), 138 (100), 100 (75), 82 (94), 56 (53).

5.5 Addition of thioamides to alkynes

5.5.1 General method

Method A: An oven-dried flask was charged with the N-nucleophile (1.00 mmol), bis(2-methallyl)-cycloocta-1,5-diene-ruthenium(II) (6.4 mg, 0.02 mmol) and molecular sieves (500 mg) and flushed with nitrogen. Subsequently, dry toluene (3.0 ml), tri-n-octylphosphine (27 µL, 0.06 mmol), alkyne (2.00 mmol) and water (108 µL, 6.00 mmol) were added via syringe. The resulting solution was stirred for 6 h at 60 °C, then poured into an aqueous sodium bicarbonate solution (30 mL). The resulting mixture was extracted repeatedly with 20 mL portions of ethyl acetate, the combined organic layers were washed with water and brine, dried with magnesium sulfate, filtered, and the volatiles were removed in vacuo. The residue was purified by column chromatography (silica gel, ethyl acetate/hexane) to yield the Z-secondary enamide. The identity and purity of the products were confirmed by ^{1}H and ^{13}C NMR spectroscopy, mass spectroscopy and elemental analysis.

Method B: An oven-dried flask was charged with the N-nucleophile (1.00 mmol), bis(2-methallyl)-cycloocta-1,5-diene-ruthenium(II) (6.4 mg, 0.02 mmol), KO^{t}Bu (5.6 mg, 0.04 mmol) and bis(dicyclohexylphosphino)methane (12.3 mg, 0.06 mmol) and flushed with nitrogen. Subsequently, dry toluene (3.0 ml) and alkyne (2.00 mmol) were added via syringe. The resulting solution was stirred for 6 h at 60 °C, then poured into an aqueous sodium bicarbonate solution (30 mL). The resulting mixture was extracted repeatedly with 20 mL portions of ethyl acetate, the combined organic layers were washed with water and brine, dried with magnesium sulfate, filtered, and the volatiles were removed in vacuo. The residue was purified by column chromatography (silica gel, ethyl acetate/hexane) to yield the Z-secondary enamide. The identity and purity of the products were confirmed by ^{1}H and ^{13}C NMR spectroscopy, mass spectroscopy and elemental analysis.

Experimental part

5.5.2 Synthesis of the thioenamides

Synthesis of N-[(E)-hex-1-enyl]pyrrolidine-2-thione (35aa).

35aa was prepared according to method A, using the following amounts: Ru(cod)[met]$_2$ (6.4 mg, 0.02 mmol), molecular sieves (500 mg), pyrrolidine-2-thione (101.2 mg, 1.00 mmol), tri-*n*-octylphosphine (27 µL, 0.06 mmol) and 1-hexyne (229 µL, 2.0 mmol). The residue was purified by column chromatography (silica gel, ethyl acetate/hexane 1:3) to afford the title product as a yellowish oil (16:1 mixture with the Z-isomer 179.7 mg, 98%).

¹H-NMR (400 MHz, CDCl$_3$, 25 °C): δ=7.40 (d, ^3J(H,H)=14.5 Hz, 1H, H-5), 5.28 (dt, ^3J(H,H)=14.5, 7.1 Hz, 1H, H-6), 3.72 (t, ^3J(H,H)=7.4 Hz, 2H, H-1), 2.96 (t, ^3J(H,H)=7.9 Hz, 2H, H-3), 2.03 (m, 4H, H-2, H-7), 1.24-1.30 (m, 4H, H-8, H-9), 0.79 (t, ^3J(H,H)=7.1 Hz, 3H, H-10) ppm.

¹³C-NMR (101 MHz, CDCl$_3$, 25 °C): δ=199.0 (C-4), 126.6 (C-5), 119.4 (C-6), 52.5 (C-7), 45.3 (C-1), 31.8 (C-8), 29.8 (C-3), 22.2 (C-9), 19.4 (C-2), 13.9 (C-10) ppm.

EA Calcd: C, 65.52; H, 9.35; N, 7.64.
Found: C, 65.09; H, 9.45; N, 7.41.

MS (Ion trap, EI): *m/z* (%) = 183 (27, [M+]), 126 (100), 98 (45), 85 (7), 45 (7).

Authentic sample of N-[(E)-hex-1-enyl]pyrrolidine-2-thione (**35aa**). An oven-dried flask was charged with Lawesson's reagent (404.5 mg, 3.00 mmol) and flushed with nitrogen. Dry toluene (7.0 ml) and a solution of N-[(E)-hex-1-enyl]pyrrolidine (836.3 mg, 5.0 mmol) in toluene (3.0 mL) were added via syringe. The resulting solution was stirred for 1 h at 110 °C, then the solvent was removed in vacuo and the residue was purified by column chromatography (silica gel, ethyl acetate/hexane 1:3) to afford product **35aa** as a yellowish oil (265.5 mg, 29%). The 1H-, 13C-NMR, and MS were correspondent to those above.

Synthesis of N-[(E)-3,3-dimethyl-but-1-enyl]pyrrolidine-2-thione (35ab)
(CAS-Nr: 99641-83-5)

35ab was prepared according to method A, using the following amounts: Ru(cod)[met]$_2$ (6.4 mg, 0.02 mmol), molecular sieves (500 mg),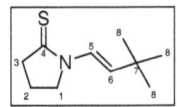

Experimental part

pyrrolidine-2-thione (101.2 mg, 1.00 mmol), tri-*n*-octylphosphine (27 µL, 0.06 mmol) and 3,3-dimethyl-1-butyne (246 µL, 2.0 mmol). The residue was purified by column chromatography (silica gel, ethyl acetate/hexane 1:3) to afford the product as a yellowish oil (13:1 mixture with the Z-isomer 177.8 mg, 97%).

^{1}H-NMR   (400 MHz, CDCl$_3$, 25 °C): δ=7.40 (d, ^3J(H,H)=14.7 Hz, 1H, H-5), 5.33 (d, ^3J(H,H)=14.7 Hz, 1H, H-5), 3.72 (t, ^3J(H,H)=7.4 Hz, 2H, H-1), 2.98 (t, ^3J(H,H)=7.8 Hz, 2H, H-3), 2.03 (tt, ^3J(H,H)=7.8, 7.6 Hz, 2H, H-2), 1.01 (s, 9H, H-8) ppm.

^{13}C-NMR   (101 MHz, CDCl$_3$, 25 °C): δ=199.4 (C-4), 130.7 (C-5), 123.5 (C-6), 52.5 (C-1), 45.4 (C-3), 32.6 (C-8), 29.9 (C-7), 19.5 (C-2) ppm.

EA   Calcd: C, 65.52; H, 9.35; N, 7.64.
Found: C, 65.23; H, 9.36; N, 7.70.

MS   (Ion trap, EI): *m/z* (%) = 184 (51, [M$^+$]), 134 (4), 128 (7), 127 (14), 126 (100).

The data correspond to those reported in the literature: E. Vedejs, R. G. Wilde, *J. Org. Chem.* **1986**, *51*, 117.

Synthesis of *N*-[(*E*)-4-phenylbut-1-en-1-yl]pyrrolidine-2-thione (35ac).

35ac was prepared according to method A, using the following amounts: Ru(cod)[met]$_2$ (6.4 mg, 0.02 mmol), molecular sieves (500 mg), pyrrolidine-2-thione (101.2 mg, 1.00 mmol), tri-*n*-octylphosphine (27 µL, 0.06 mmol) and 4-phenyl-1-butyne (281 µL, 2.0 mmol). The residue was purified by column chromatography (silica gel, ethyl acetate/hexane 2:3) to afford the product as a yellowish oil (7:1 mixture with the Z-isomer, 219.8 mg, 95%).

^{1}H-NMR   (400 MHz, CDCl$_3$, 25 °C): δ=7.63 (d, ^3J(H,H)=14.3 Hz, 1H, H-5), 7.33 (m, 2H, H$_{arom}$), 7.23-7.24 (m, 3H, H$_{arom}$), 5.42 (dt, ^3J(H,H)=16.3, 7.1 Hz, 1H, H-6), 3.79 (t, ^3J(H,H)=7.4 Hz, 2H, H-1), 3.10 (t, ^3J(H,H)=7.8 Hz, 2H, H-3), 2.79 (t, ^3J(H,H)=7.6 Hz, 2H, H-8), 2.50-2.55 (m, 2H, H-7), 2.07-2.13 (m, 2H, H-2) ppm.

Experimental part

13C-NMR (101 MHz, CDCl$_3$, 25 °C): δ=199.0 (C-4), 140.7 (C-5), 128.0 (C$_{arom}$, double peak), 126.6 (C$_{arom}$), 125.6 (C$_{arom}$), 117.8 (C-6), 52.0 (C-1), 44.9 (C-3), 35.8 (C-8), 31.6 (C-7), 18.9 (C-2) ppm.

EA Calcd: C, 72.68; H, 7.41; N, 6.05.
Found: C, 72.37; H, 7.60; N, 5.90.

MS (Ion trap, EI): *m/z* (%) = 231 (2, [M$^+$]), 126 (100), 98 (10), 58 (4), 45 (4).

Synthesis of *N*-[(*E*)-5-chloro-pent-1-enyl]pyrrolidine-2-thione (35ad)

35ad was prepared according to method A, using the following amounts: Ru(cod)[met]$_2$ (16 mg, 0.05 mmol), molecular sieves (500 mg) and pyrrolidine-2-thione (101.2 mg, 1.00 mmol), tri-*n*- 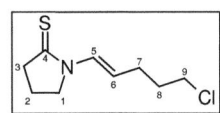 octylphosphine (67 μL, 0.15 mmol) and 5-chloro-1-pentyne (210 μL, 2.0 mmol). The residue was purified by column chromatography (silica gel, ethyl acetate/hexane 1:3) to afford the product as a yellowish oil (15:1 mixture with the *Z*-isomer, 161,1 mg, 78%).

1H-NMR (400 MHz, CDCl$_3$, 25 °C): δ=7.51 (d, ^3J(H,H)=14.5 Hz, 1H, H-5), 5.28 (dt, ^3J(H,H)=14.5, 7.2 Hz, 1H, H-6), 3.74-3.79 (m, 2H, H-1), 3.47-3.53 (m, 2H, H-9), 3.02 (t, ^3J(H,H)=7.9 Hz, 2H, H-3), 2.24-2.30 (m, 2H, H-7), 2.05-2.11 (m, 2H, H-2), 1.82-1.88 (m, 2H, H-7) ppm.

13C-NMR (101 MHz, CDCl$_3$, 25 °C): δ=199.5 (C-4), 127.4 (C-5), 116.8 (C-6), 52.2 (C-1), 45.1 (C-3), 44.0 (C-9), 32.2 (C-7), 27.2 (C-8), 19.3 (C-2) ppm.

EA Calcd: C, 53.06; H, 6.93; N, 6.87.
Found: C, 53.42; H, 6.60; N, 7.09.

MS (Ion trap, EI): *m/z* (%) = 203 (6, [M$^+$]), 168 (9), 126 (100), 98 (8), 45 (6).

Experimental part

Synthesis of N-[(E)-hept-1-en-6-ynyl]pyrrolidine-2-thione (35ae)

35ae was prepared according to method A, using the following amounts: Ru(cod)[met]$_2$ (6.4 mg, 0.02 mmol), molecular sieves (500 mg), pyrrolidine-2-thione (101.2 mg, 1.00 mmol), tri-n-octylphosphine (27 µL, 0.06 mmol) and 1,6-heptadiyne (229 µL, 2.0 mmol). The residue was purified by column chromatography (silica gel, ethyl acetate/hexane 1:3) to afford the product as a yellowish oil (11:1 mixture with the Z-isomer, 146.9 mg, 76%).

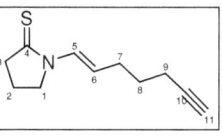

^{1}H-NMR	(400 MHz, CDCl$_3$, 25 °C): δ=7.50 (d, ^3J(H,H)=14.5 Hz, 1H, H-5), 5.30 (dt, ^3J(H,H)=14.5, 7.3 Hz, 1H, H-6), 3.77 (t, ^3J(H,H)=7.5 Hz, 2H, H-1), 3.03 (t, ^3J(H,H)=8.0 Hz, 2H, H-3), 2.00-2.29 (m, 6H, H-2, H-7, H-9), 1.92 (t, ^3J(H,H)=2.7 Hz, 1H, H-11), 1.53-1.68 (m, 2H, H-8) ppm.
^{13}C-NMR	(101 MHz, CDCl$_3$, 25 °C): δ=199.4 (C-4), 127.2 (C-5), 117.7 (C-6), 83.7 (C-10), 68.7 (C-11), 52.3 (C-1), 45.1 (C-3), 28.9 (C-7), 28.3 (C-9), 19.3 (C-2), 17.7 (C-8) ppm.
EA	Calcd: C, 68.35; H, 7.82; N, 7.25. Found: C, 68.47; H, 7.93; N, 7.21.
MS	(Ion trap, EI): m/z (%) = 193 (100, [M$^+$]), 165 (10), 166 (33), 140 (4), 126 (55).

Synthesis of N-[(E)-3-methoxy-propenyl]pyrrolidine-2-thione (35af)

35af was prepared according to method A, using the following amounts: Ru(cod)[met]$_2$ (16 mg, 0.05 mmol), molecular sieves (500 mg), pyrrolidine-2-thione (101.2 mg, 1.00 mmol), tri-n-octylphosphine (67 µL, 0.15 mmol), 3-methoxy-1-propyne (169 µL, 2.0 mmol). The residue was purified by column chromatography (silica gel, ethyl acetate/hexane 1:3) to afford the product as a yellowish oil (11:1 mixture with the Z-isomer, 130.2 mg, 76%).

^{1}H-NMR	(400 MHz, CDCl$_3$, 25 °C): δ=7.66 (d, ^3J(H,H)=14.5 Hz, 1H, H-5), 5.37 (dt, ^3J(H,H)=14.5, 6.7 Hz, 1H, H-7), 3.98 (d, ^3J(H,H)=6.7 Hz, 2H, H-7), 3.79 (t, ^3J(H,H)=7.4 Hz, 2H, H-1), 3.29 (s, 3H, H-8), 3.04 (t, ^3J(H,H)=7.8 Hz, 2H, H-3), 2.07-2.12 (m, 2H, H-2) ppm.

¹³C-NMR	(101 MHz, CDCl₃, 25 °C): δ=201.5 (C-4), 129.4 (C-5), 113.8 (C-6), 70.9 (C-7), 57.9 (C-8), 52.3 (C-1), 45.4 (C-3), 19.5 (C-1) ppm.
EA	Calcd: C, 56.11; H, 7.65; N, 8.18. Found: C, 56.29; H, 7.93; N, 7.87.
MS	(Ion trap, EI): *m/z* (%) = 171 (5, [M⁺]), 140 (40), 126 (100), 85 (15), 45 (17).

Synthesis of N-[(E)-3-trimethylsilanyloxypropenyl]pyrrolidine-2-thione (35ag)

35ag was prepared according to method A, using the following amounts: Ru(cod)[met]₂ (6.4 mg, 0.02 mmol), molecular sieves (500 mg), pyrrolidine-2-thione (101.2 mg, 1.00 mmol), tri-*n*-octylphosphine (27 µL, 0.06 mmol) and trimethyl(propargyloxy)silane (308 µL, 2.0 mmol).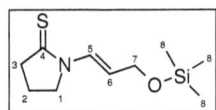
The residue was purified by column chromatography (silica gel, ethyl acetate/hexane 3:2) to afford the title product as a yellow solid (19:1 mixture with the Z-isomer, 130.8 mg, 57%).

¹H-NMR	(400 MHz, CDCl₃, 25 °C): δ=7.66 (d, ³*J*(H,H)=14.5 Hz, 1H, H-5), 5.41 (dt, ³*J*(H,H)=14.5, 6.2 Hz, 1H, H-5), 4.2 (d, ³*J*(H,H)=6.2 Hz, 2H, H-7), 3.79 (t, ³*J*(H,H)= 7.4 Hz, 2H, H-1), 3.03 (t, ³*J*(H,H)=7.8 Hz, 2H, H-3), 2.06-2.11 (m, 2H, H-2), 0.10 (s, 9H, H-8) ppm.
¹³C-NMR	(101 MHz, CDCl₃, 25 °C): δ=201.1 (C-4), 127.8 (C-5), 116.8 (C-6), 61.4 (C-7), 52.3 (C-1), 45.4 (C-3), 19.5 (C-2), 0.3 (C-8) ppm.
EA	Calcd: C, 52.35; H, 8.35; N, 6.11. Found: C, 52.59; H, 8.56; N, 6.01.
MS	(Ion trap, EI): *m/z* (%) = 229 (3, [M⁺]), 158 (10), 140 (10), 126 (100), 45 (9).

Synthesis of N-[(E)-trimethylsilanyl-vinyl]pyrrolidine-2-thione (35ah)

35ah was prepared according to method A, using the following amounts: Ru(cod)[met]₂ (16 mg, 0.05 mmol), molecular sieves (500 mg) and pyrrolidine-2-thione (101.2 mg, 1.00 mmol), tri-*n*-octylphosphine (67 µL,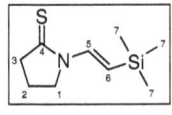
0.15 mmol) and trimethylsilylacetylene (282 µL, 2.0 mmol). The residue was purified by column chromatography (silica gel, ethyl acetate/hexane 1:3) to afford the title product as a white solid (1:1 mixture with the Z-isomer, 175.5 mg, 88%).

Experimental part

¹H-NMR (400 MHz, CDCl₃, 25 °C): δ=7.71 (d, ³J(H,H)=17.2 Hz, 1H, H-5), 5.20 (d, ³J(H,H)=17.2 Hz, 1H, H-6), 3.82 (t, ³J(H,H)=7.4 Hz, 2H, H-1), 3.06 (t, ³J(H,H)=7.8 Hz, 2H, H-3), 2.10 (m, 2H, H-2), 0.17 (s, 9H, H-7) ppm.

¹³C-NMR (101 MHz, CDCl₃, 25 °C): δ=201.1 (C-4), 135.9 (C-5), 113.7 (C-6), 51.6 (C-1), 45.0 (C-3), 19.3 (C-2), 0.9 (C-7) ppm.

EA Calcd: C, 54.21; H, 8.59; N, 7.02.
Found: C, 54.38; H, 8.50; N, 7.29.

MS (Ion trap, EI): *m/z* (%) = 199 (3, [M⁺]), 184 (12), 158 (8), 126 (100), 98 (11).

Synthesis of N-[(E)-3-methyl-buta-1,3-dienyl]pyrrolidine-2-thione (35ai)

35ai was prepared according to method A, using the following amounts: Ru(cod)[met]₂ (16 mg, 0.05 mmol), molecular sieves (500 mg), pyrrolidine-2-thione (101.2 mg, 1.00 mmol), tri-*n*-octylphosphine (67 µL, 0.15 mmol), 2-methyl-1-buten-3-yne (192 µL, 2.0 mmol). The residue was purified by column chromatography (silica gel, ethyl acetate/hexane 1:3) to afford the product as a yellow solid (3:1 mixture with the Z-isomer, 118.8 mg, 71%).

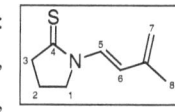

¹H-NMR (400 MHz, CDCl₃, 25 °C): δ=7.70 (d, ³J(H,H)=14.7 Hz, 1H, H-5), 6.05 (d, ³J(H,H)=14.7 Hz, 1H, H-6), 5.03 (d, 2H, H-7), 3.87 (m, 2H, H-1), 3.11 (m, 2H, H-3), 2.14 (m, 2H, H-2), 1.94 (s, 3H, H-8) ppm.

¹³C-NMR (101 MHz, CDCl₃, 25 °C): δ=200.5 (C-4), 140.3 (C-5), 126.3 (C-7), 120.9 (C-7), 117.3 (C-6), 52.3 (C-1), 45.3 (C-3), 19.5 (C-2), 18.7 (C-8) ppm.

EA Calcd: C, 64.62; H, 7.83; N, 8.37.
Found: C, 64.43; H, 7.83; N, 7.55.

MS (Ion trap, EI): *m/z* (%) = 167 (41, [M⁺]), 134 (100), 126 (93), 98 (17), 58 (11).

Synthesis of N-[(E)-2-phenylvinyl]pyrrolidine-2-thione (35aj)

35aj was prepared according to method A, using the following amounts: Ru(cod)[met]₂ (6.4 mg, 0.02 mmol), molecular sieves (500 mg), pyrrolidine-2-thione (101.2 mg, 1.00 mmol), tri-*n*-octylphosphine (27 µL, 0.06 mmol) and phenylacetylene (220 µL, 2.0 mmol). The residue

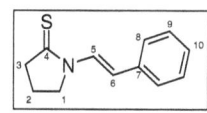

was purified by column chromatography (silica gel, ethyl acetate/hexane 1:4) to afford the product as a yellow solid (6:1 mixture with the Z-isomer, 197.29 mg, 97%).

¹H-NMR (400 MHz, CDCl$_3$, 25 °C): δ=8.24 (d, ^3J(H,H)=14.9 Hz, 1H, H-5), 7.41 (m, 2H, H$_{arom}$), 7.30 (m, 2H, H$_{arom}$), 7.22 (m, 1H, H$_{arom}$), 6.21 (d, ^3J(H,H)=14.9 Hz, 1H, H-6), 3.91 (m, 2H, H-1), 3.09 (m, 2H, H-3), 2.13 (m, 2H, H-2) ppm.

¹³C-NMR (101 MHz, CDCl$_3$, 25 °C): δ=200.5 (C-4), 135.2 (C-5), 128.6 (C$_{arom}$), 127.4 (C$_{arom}$), 126.1 (C$_{arom}$), 125.9 (C$_{arom}$), 117.6 (C-6), 52.2 (C-1), 45.1 (C-3), 19.3 (C-2) ppm.

EA Calcd: C, 70.89; H, 6.45; N, 6.89.
Found: C, 71.02; H, 6.65; N, 6.57.

MS (Ion trap, EI): m/z (%) = 203 (74, [M⁺]), 174 (6), 126 (100), 77 (16), 44 (12).

Synthesis of N-[(E)-2-phenylvinyl]piperidine-2-thione (35bj)

35bj was prepared according to method A, using the following amounts: Ru(cod)[met]$_2$ (16 mg, 0.05 mmol), molecular sieves (500 mg), piperidine-2-thione (115.2 mg, 1.00 mmol), tri-n-octylphosphine (67 µL, 0.15 mmol) and phenylacetylene (220 µL, 2.0 mmol). The residue was purified by column chromatography (silica gel, ethyl acetate/hexane 1:9) to afford the product as a yellow solid (5:1 mixture with the Z-isomer, 193.4 mg, 89%).

¹H-NMR (400 MHz, CDCl$_3$, 25 °C): δ=8.87 (d, ^3J(H,H)=14.9 Hz, 1H, H-6), 7.41 (d, 2H, H$_{arom}$), 7.31 (t, 2H, H$_{arom}$), 7.23 (d, 1H, H$_{arom}$), 6.38 (d, ^3J(H,H)=14.9 Hz, 1H, H-7), 3.72 (m, 2H, H-1), 3.14 (m, 2H, H-4), 2.00-2.04 (m, 2H, H-2), 1.76-1.80 (m, 2H, H-3) ppm.

¹³C-NMR (101 MHz, CDCl$_3$, 25 °C): δ=201.1 (C-5), 131.9 (C-6), 128.8 (C$_{arom}$), 128.6 (C$_{arom}$), 127.5 (C$_{arom}$), 126.4 (C$_{arom}$), 117.2 (C-7), 48.4 (C-1), 42.9 (C-4), 22.6 (C-2), 20.1 (C-3) ppm.

EA Calcd: C, 71.85; H, 6.96; N, 6.44.
Found: C, 71.96; H, 7.26; N, 6.41.

MS (Ion trap, EI): m/z (%) = 217 (54, [M⁺]), 174 (6), 140 (100), 98 (11), 51 (16).

Synthesis of N-[(E)-2-phenylvinyl]azepane-2-thione (35cj)

35cj was prepared according to method A, using the following amounts: Ru(cod)[met]$_2$ (16 mg, 0.05 mmol), molecular sieves (500 mg), 2H-azepine-2-thione (129.2 mg, 1.00 mmol), tri-n-octylphosphine (67 μL, 0.15 mmol) and phenylacetylene (220 μL, 2.0 mmol). The residue was purified by column chromatography (silica gel, ethyl acetate/hexane 1:4) to afford the product as a yellow solid (30:1 mixture with the Z-isomer, 194.3 mg, 84%).

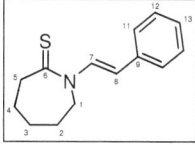

¹H-NMR (400 MHz, CDCl$_3$, 25 °C): δ=8.59 (d, ^3J(H,H)=14.7 Hz, 1H, H-7), 7.40 (d, 2H, H$_{arom}$), 7.31 (t, 2H, H$_{arom}$), 7.22 (t, 1H, H$_{arom}$), 6.31 (d, ^3J(H,H)=14.7 Hz, 1H, H-8), 4.02 (m, 2H, H-1), 3.28 (m, 2H, H-5), 1.78 (m, 6H, H-2, 3, 4) ppm.

¹³C-NMR (101 MHz, CDCl$_3$, 25 °C): δ=206.3 (C-6), 135.7 (C-7), 132.8 (C$_{arom}$), 128.7 (C$_{arom}$), 127.4 (C$_{arom}$), 126.3 (C$_{arom}$), 117.1 (C-8), 50.5 (C-1), 47.5 (C-5), 28.6 (C-2), 26.6 (C-4), 25.0 (C-3) ppm.

EA Calcd: C, 72.68; H, 7.41; N, 6.05.
Found: C, 73.08; H, 7.28; N, 5.80.

MS (Ion trap, EI): m/z (%) = 231 (68, [M⁺]), 154 (100), 96 (31), 69 (12), 45 (9).

Synthesis of N-phenyl-N-[(E)-2-phenylvinyl]thioacetamide (35dj)

35dj was prepared according to method A, using the following amounts: Ru(cod)[met]$_2$ (16 mg, 0.05 mmol), molecular sieves (500 mg), N-phenyl-thioacetamide (151.2 mg, 1.00 mmol), tri-n-octylphosphine (67 μL, 0.15 mmol) and phenylacetylene (220 μL, 2.0 mmol). The residue was purified by column chromatography (silica gel, ethyl acetate/hexane 1:4) to afford the product as a yellow solid (30:1 mixture with the Z-isomer, 248.3 mg, 98%).

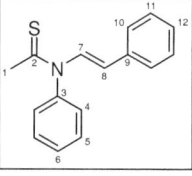

¹H-NMR (400 MHz, CDCl$_3$, 25 °C): δ=9.01 (d, ^3J(H,H)=14.67 Hz, 1H, H-7), 7.50-7.57 (m, 3H, H$_{arom}$), 7.16-7.29 (m, 7H, H$_{arom}$), 5.56 (d, ^3J(H,H)=14.67 Hz, 1H, H-8), 2.39 (s, 3H, H-1) ppm.

Experimental part

¹³C-NMR (101 MHz, CDCl₃, 25 °C): δ=200.4 (C-2), 141.2 (C$_{arom}$), 135.6 (C-7), 133.2 (C$_{arom}$), 130.2 (C$_{arom}$), 129.2 (C$_{arom}$), 128.6 (C$_{arom}$), 127.5 (C$_{arom}$), 127.4 (C$_{arom}$), 126.3 (C$_{arom}$), 120.1 (C-8), 35.0 (C-1) ppm.

EA Calcd: C, 75.85; H, 5.97; N, 5.53.
Found: C, 76.07; H, 6.23; N, 5.66.

MS (Ion trap, EI): m/z (%) = 253(18, [M⁺]), 176 (5), 118 (100), 77 (14), 59 (10).

Synthesis of N-phenyl-N-[(E)-2-phenylvinyl]thiobenzamide (35ej)

35ej was prepared according to method A, using the following amounts: Ru(cod)[met]₂ (6.4 mg, 0.02 mmol), molecular sieves (500 mg), N-phenyl-thiobenzamide (231.3 mg, 1.00 mmol), tri-n-octylphosphine (27 µL, 0.06 mmol) and phenylacetylene (220 µL, 2.0 mmol). The residue was purified by column 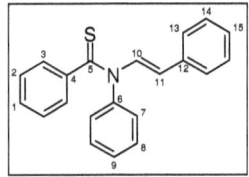 chromatography (silica gel, ethyl acetate/hexane 1:3) to afford the product as a yellow solid (30:1 mixture with the Z-isomer, 306.0 mg, 97%).

¹H-NMR (400 MHz, CDCl₃, 25 °C): δ=7.23-7.33 (m, 16H, H$_{arom}$), 5.87 (d, 1H, H-11) ppm.

¹³C-NMR (101 MHz, CDCl₃, 25 °C): δ=201.6 (C-5), 143.5 (C$_{arom}$), 142.3 (C$_{arom}$), 135.5 (C-10), 133.8 (C$_{arom}$), 129.4 (C$_{arom}$), 128.8 (C$_{arom}$), 128.7 (C$_{arom}$), 128.4 (C$_{arom}$), 128.2 (C$_{arom}$), 127.9 (C$_{arom}$), 127.7 (C$_{arom}$), 127.5 (C$_{arom}$), 126.2 (C-11) ppm.

EA Calcd: C, 79.96; H, 5.43; N, 4.44.
Found: C, 79.67; H, 5.31; N, 4.44.

MS (Ion trap, EI): m/z (%) = 315 (10, [M⁺]), 180 (100), 121 (29), 77 (23), 51 (12).

Synthesis of N-methyl-N-[(E)-2-phenylvinyl]thiobenzamide (35fj)

35fj was prepared according to method A, using the following amounts: Ru(cod)[met]₂ (16 mg, 0.05 mmol), molecular sieves (500 mg), N-methyl-thiobenzamide (151.2 mg, 1.00 mmol), tri-n-octylphosphine (67 µL, 0.15 mmol) and phenylacetylene

(220 μL, 2.0 mmol). The residue was purified by column chromatography (silica gel, ethyl acetate/hexane 1:4) to afford the product as a yellow solid (15:1 mixture with the Z-isomer, 230.6 mg, 91%).

¹H-NMR (400 MHz, CDCl₃, 25 °C): δ=7.12-7.52 (m, 11H, H$_{arom}$), 6.25 (d, 1H, H-8), 3.97 (s, 3H, H-6), ppm.

¹³C-NMR (101 MHz, CDCl₃, 25 °C): δ=201.5 (C-5), 142.9 (C$_{arom}$), 135.4 (C-7), 131.3 (C$_{arom}$), 129.3 (C$_{arom}$), 128.8 (C$_{arom}$), 128.4 (C$_{arom}$), 127.3 (C$_{arom}$), 127.04 (C$_{arom}$), 125.8 (C$_{arom}$), 114.8 (C-8), 37.8 (C-6) ppm.

EA Calcd: C, 75.85; H, 5.97; N, 5.53.
Found: C, 76.12; H, 6.19; N, 5.63.

MS (Ion trap, EI): m/z (%) = 253 (49, [M⁺]), 176 (50), 121 (75), 118 (100), 77 (48).

Synthesis of N-[(E)-2-phenylvinyl]-1,3-oxazinane-2-thione (35gj)

35gj was prepared according to method A, using the following amounts: Ru(cod)[met]₂ (16 mg, 0.05 mmol), molecular sieves (500 mg), 1,3-oxazinane-2-thione (117.2 mg, 1.00 mmol), tri-n-octylphosphine (67 μL, 0.15 mmol) and phenylacetylene (220 μL, 2.0 mmol). The residue was purified by column chromatography (silica gel, ethyl acetate/hexane 1:3) to afford the product as a yellow solid (2:1 mixture with the Z-isomer, 92.1 mg, 42%).

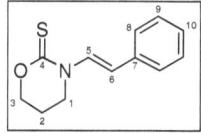

¹H-NMR (400 MHz, CDCl₃, 25 °C): δ=8.05 (d, ³J(H,H)=15.0 Hz, 1H, H-6), 7.26-7.35 (m, 4H, H$_{arom}$), 7.18 (t, 1H, H$_{arom}$), 5.95 (d, ³J(H,H)=15.0 Hz, 1H, H-6), 3.71 (t, 2H, H-1), 3.10 (t, 2H, H-3), 2.33 (tt, 2H, H-2) ppm.

¹³C-NMR (101 MHz, CDCl₃, 25 °C): δ=164.6 (C-4), 136.4 (C-5), 128.7 (C$_{arom}$), 127.3 (C$_{arom}$), 126.7 (C$_{arom}$), 125.9 (C$_{arom}$), 110.7 (C-6), 45.8 (C-1), 28.3 (C-3), 23.8 (C-2) ppm.

EA Calcd: C, 65.72; H, 5.97; N, 6.39.
Found: C, 66.08; H, 6.32; N, 6.27.

MS (Ion trap, EI): m/z (%) = 219 (100, [M⁺]), 191 (35), 130 (54), 103 (19), 77 (18).

Experimental part

Synthesis of N-[(Z)-hex-1-enyl]pyrrolidine-2-thione (36aa)

36aa was prepared according to method B, using the following amounts: Ru(cod)[met]$_2$ (6.4 g, 0.05 mmol), KO^{t}Bu (5.6 mg, 0.04 mmol), pyrrolidine-2-thione (101.2 mg, 1.00 mmol), bis(dicyclohexylphosphino)methane (12.3 mg, 0.06 mmol) and 1-hexyne (229 μL, 2.0 mmol). The residue was purified by column chromatography (silica gel, ethyl acetate/hexane 1:3) to afford the product as a yellowish oil (1:3 mixture with the E-isomer, 135 mg, 74%).

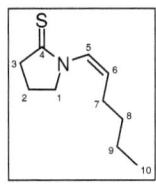

^{1}H-NMR (400 MHz, CDCl$_3$, 25 °C): δ=6.72 (d, ^3J(H,H)=9.5 Hz, 1H, H-5), 5.08-5.24 (dt, ^3J(H,H)=9.5 Hz, 1H, H-6), 3.97 (t, 2H, H-1), 3.00 (t, 2H, H-3), 2.02-2.21 (m, 4H, H-7, 2), 1.24-1.44 (m, 4H, H-8, 9), 0.87 (t, 3H, H-10) ppm.

^{13}C-NMR (101 MHz, CDCl$_3$, 25 °C): δ=202.3 (C-4), 125.8, 123.3, 56.2 (C-7), 44.3 (C-1), 31.7 (C-8), 27.2 (C-3), 22.2 (C-9), 21.1 (C-2), 13.8 (C-10) ppm.

EA Calcd: C, 65.52; H, 9.35; N, 7.64.
Found: C, 65.49; H, 9.23; N, 7.68.

MS (Ion trap, EI): m/z (%) = 183 (58, [M+]), 126 (100), 98 (13), 85 (28), 45 (10).

Synthesis of N-[(Z)-3,3-dimethyl-but-1-enyl]pyrrolidine-2-thione (36ab)

36ab was prepared according to method B, using the following amounts: Ru(cod)[met]$_2$ (6.4 g, 0.05 mmol), KO^{t}Bu (5.6 mg, 0.04 mmol), pyrrolidine-2-thione (101.2 mg, 1.00 mmol), bis(dicyclohexylphosphino)methane (12.3 mg, 0.06 mmol) and 3,3-dimethyl-1-butyne (246 μL, 2.0 mmol). The residue was purified by column chromatography (silica gel, ethyl acetate/hexane 1:3) to afford the product as a yellowish oil (1:8 mixture with the E-isomer, 128.32 mg, 70%).

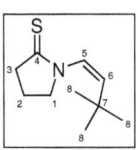

^{1}H-NMR (400 MHz, CDCl$_3$, 25 °C): δ=5.77 (d, ^3J(H,H)=9.4 Hz, 1H, H-5), 5.41 (d, ^3J(H,H)=9.4 Hz, 1H, H-6), 3.72 (t, 2H, H-1), 2.94 (t, 2H, H-3), 2.05-2.11 (m, 2H, H-2), 1.05 (s, 9H, H-8) ppm.

^{13}C-NMR (101 MHz, CDCl$_3$, 25 °C): δ=203.1 (C-4), 141.4 (C-5), 123.1 (C-6), 58.3 (C-1), 44.5 (C-3), 33.5 (C-7), 29.2 (C-8), 20.7 (C-2) ppm.

EA Calcd: C, 65.52; H, 9.35; N, 7.64.
Found: C, 65.27; H, 9.55; N, 7.65.

Experimental part

MS (Ion trap, EI): m/z (%) = 183 (55, [M⁺]), 134 (11), 128 (5), 127 (35), 126 (100).

Synthesis of N-[(Z)-4-phenylbut-1-en-1-yl]pyrrolidine-2-thione (36ac)

36ac was prepared according to method B, using the following amounts: Ru(cod)[met]₂ (6.4 mg, 0.05 mmol), KO^{t}Bu (5.6 mg, 0.04 mmol), pyrrolidine-2-thione (101.2 mg, 1.00 mmol), bis(dicyclohexylphosphino)methane (12.3 mg, 0.06 mmol) and 4-phenyl-1-butyne (281 μL, 2.0 mmol). The residue was purified by column chromatography (silica gel, ethyl acetate/hexane 2:3) to afford the product as a yellowish oil (1:2 mixture with the E-isomer, 128.3 mg, 64%).

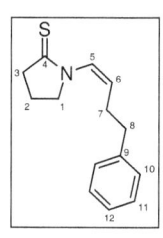

¹H-NMR	(400 MHz, CDCl₃, 25 °C): δ=7.33 (d, ^3J(H,H)=7.6 Hz, 2H, H$_{arom}$), 7.16-7.28 (m, 3H, H$_{arom}$), 6.82 (d, ^3J(H,H)=9.5 Hz, 1H, H-5), 5.23-5.35 (dt, ^3J(H,H)=9.5 Hz, 1H, H-6), 3.86 (t, 2H, H-1), 3.02 (t, 2H, H-3), 2.72-2.84 (m, 4H, H-7, 8), 2.11 (m, 2H, H-2) ppm.
¹³C-NMR	(101 MHz, CDCl₃, 25 °C): δ=199.5 (C-4), 141.2 (C-5), 128.54 (C$_{arom}$), 128.50 (C$_{arom}$), 127.1 (C$_{arom}$), 126.1 (C$_{arom}$), 121.9 (C-6), 56.2 (C-1), 45.3 (C-3), 35.7 (C-7), 32.1 (C-8), 21.2 (C-2) ppm.
EA	Calcd: C, 72.68; H, 7.41; N, 6.05. Found: C, 72.78; H, 7.52; N, 5.95.
MS	(Ion trap, EI): m/z (%) = 231 (22, [M⁺]), 126 (100), 98 (8), 58 (5), 45 (10).

5.6 Markovnikov addition of secondary amides to alkynes

Synthesis of 1-(1-phenyl-vinyl)-pyrrolidin-2-one (47aa)

An oven-dried flask was charged flushed with nitrogen. Subsequently, dry toluene (3.0 mL), tri-n-butyl phosphine (22 µL, 0.09 mmol), pyrolidinone (1 mmol) and 1-hexyne (2.00 mmol) were added via syringe. 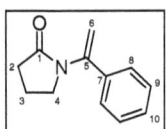 The resulting solution was stirred for 15 h at 100 °C, and then poured into an aqueous sodium bicarbonate solution (30 mL). The resulting mixture was extracted repeatedly with 20 mL portions of ethyl acetate, the combined organic layers were washed with water and brine, dried with magnesium sulfate, filtered, and the volatiles were removed in vacuo. The residue was purified by column chromatography (silica gel, ethyl acetate/hexane) to yield the enamide **47aa** (159.0 mg, 85%). The identity and purity of the products were confirmed by ^{1}H and ^{13}C NMR spectroscopy, and mass.

^{1}H-NMR (400 MHz, CDCl$_3$, 25 °C): δ=7.25-7.29 (m, 5H, H$_{arom}$), 5.33 (s, 1H, H-6), 5.23 (s, 1H, H-6), 3.47 (t, ^3J(H,H)=7.0 Hz, 2H, H-4), 2.49 (t, ^3J(H,H)=8.1 Hz, 2H, H-2), 2.04 (dt, ^3J(H,H)=7.2, 7.9, H-3) ppm.

^{13}C-NMR (101 MHz, CDCl$_3$, 25 °C): δ=174.2 (C-1), 143.2 (C-5), 135.9 (C$_{arom}$), 128.08 (C$_{arom}$), 128.04 (C$_{arom}$), 126.0 (C$_{arom}$), 108.9 (C-6), 49.2 (C-4), 31.5 (C-2), 18.2 (C-1) ppm.

MS (Ion trap, EI): m/z (%) = 187 (100, [M$^+$]), 159 (22), 132 (39), 77 (9), 51 (7).

Experimental part

5.7 Procedure for the mechanistic studies

5.7.1 ESI-MS experiments

Experiments with [Ru(cod)(met)$_2$]: An oven-dried flask was charged with [Ru(cod)(met)$_2$] (6.4 mg, 0.02 mmol) and 4-dimethylaminopyridine (4.9 mg, 0.04 mmol) and the flask was flushed with nitrogen. Subsequently, dry toluene (3.0 mL), pyrrolidin-2-one (1a, 77 µL, 1.00 mmol) and tri-n-butylphosphine (15 µL, 0.06 mmol) were added via syringe. The resulting solution was stirred for 1 h at 100 °C. A sample of the reaction mixture was taken for ESI-MS experiment (Figure 6, Figure 7 and Figure 8).

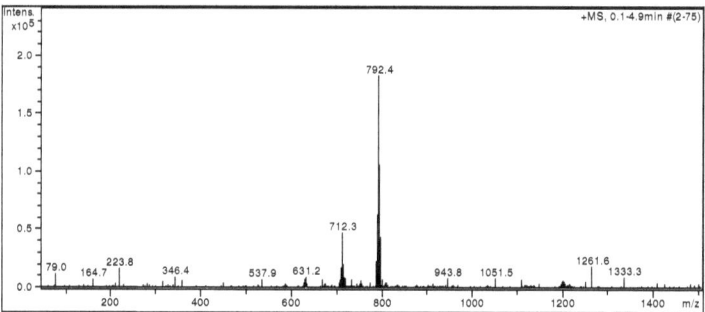

Figure 6: ESI-MS experiment with [Ru(cod)(met)$_2$].

Experimental part

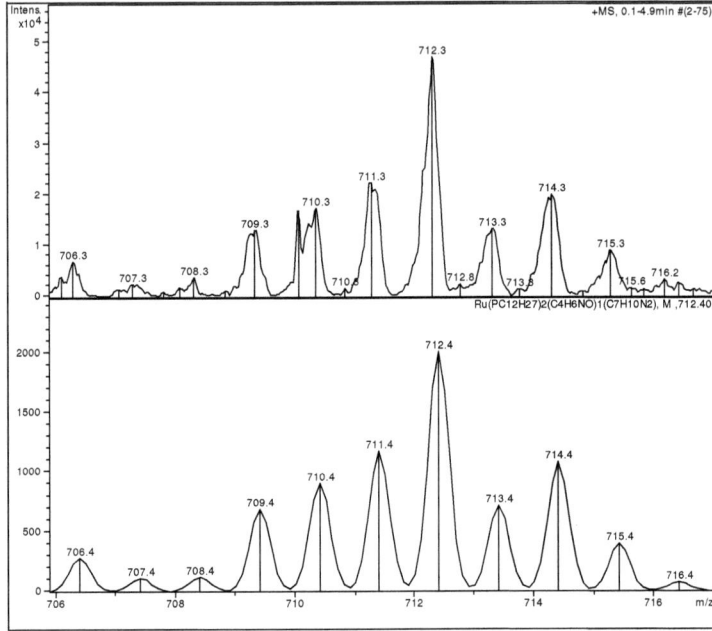

m/z (% relative to peak at 712.3) = 706.3 (14), 707.3 (5), 708.3 (7), 709.3 (27), 710.4 (36), 711.3 (47), 712.3 (100), 713.3 (27), 714.3 (42), 715.3 (18), 716.2 (7), Calc.: 706.4 (14), 707.4 (6), 708.4 (6), 709.4 (34), 710.4 (45), 711.4 (59), 712.4 (100), 713.4 (36), 714.4 (54), 715.4 (20), 716.4 (4).

Figure 7: amplified area of the signal at m/z = 712.3 and simulated spectra of

$[(n\text{-}Bu_3P)_2Ru(C_6H_6NO)(C_7H_{10}N_2)]^+$.

Experimental part

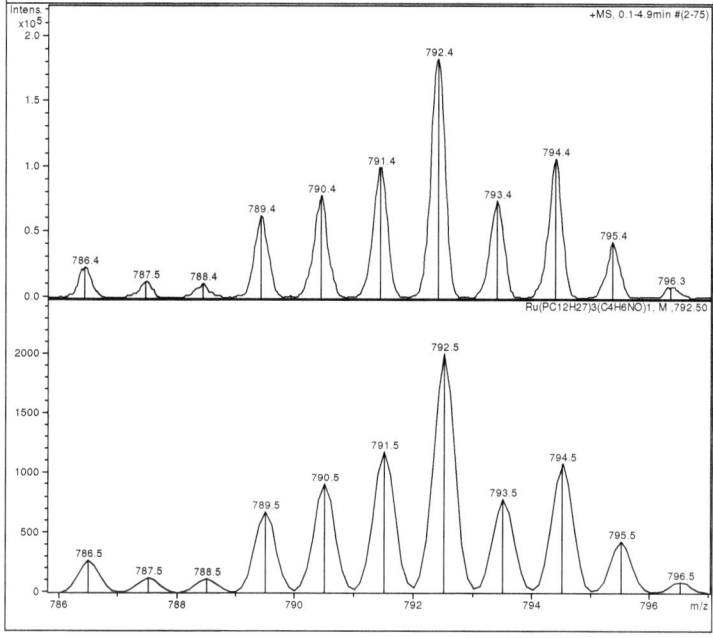

m/z (% relative to peak at 792.4) = 786.4 (13), 787.5 (7), 788.4 (6), 789.4 (34), 790.4 (43), 791.4 (55), 792.4 (100), 793.4 (40), 794.4 (58), 795.4 (24), 796.3 (5), Calc.: 786.5 (14), 787.5 (6), 788.6 (6), 789.5 (34), 790.5 (46), 791.5 (59), 792.5 (100), 793.5 (39), 794.5 (54), 795.5 (22), 796.5 (5).

Figure 8: amplified area of the signal at m/z = 792.4 and simulated spectra of $[(n\text{-Bu}_3P)_3Ru(C_6H_6NO)]^+$.

Experimental part

Experiments with RuCl₃·3 H₂O: An oven-dried flask was charged with potassium carbonate (13.8 mg, 0.10 mmol), and a stock solution containing ruthenium(III) chloride hydrate (5.2 mg, 0.02 mmol), 4-dimethylaminopyridine (4.9 mg, 0.04 mmol) and acetone (1.0 mL). The acetone was removed in vacuo and the flask was flushed with nitrogen. Subsequently, dry toluene (3.0 mL), water (5 µL, 0.30 mmol), pyrrolidin-2-one (1a, 77 µL, 1.00 mmol) and tri-n-butylphosphine (15 µL, 0.06 mmol) were added via syringe. The resulting solution was stirred for 1 h at 100 °C. A sample of the reaction mixture was taken for ESI-MS experiment.

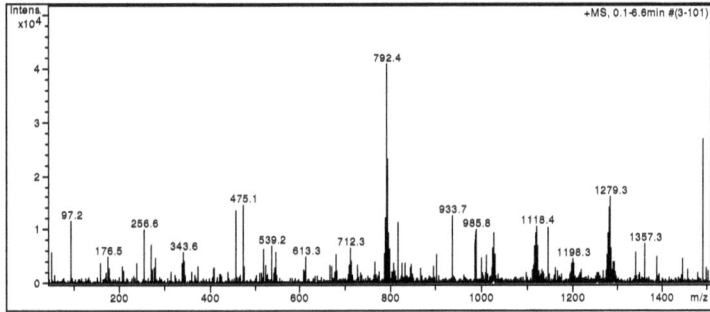

Figure 9: ESI-MS experiment with RuCl₃·3 H₂O.

Experimental part

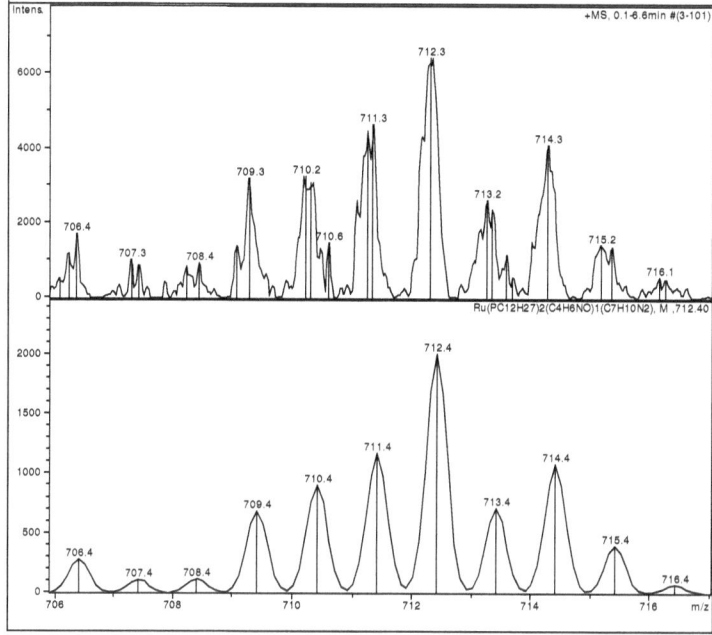

m/z (% relative to peak at 712.3) = 706.4 (26), 707.3 (15), 708.4 (14), 709.3 (50), 710.2 (51), 711.3 (74), 712.3 (100), 713.3 (42), 714.3 (63), 715.2 (24), 716.1 (10), Calc.: 706.4 (14), 707.4 (6), 708.4 (6), 709.4 (34), 710.4 (45), 711.4 (59), 712.4 (100), 713.4 (36), 714.4 (54), 715.4 (20), 716.4 (4).

Figure 10: amplified area of the signal at m/z = 712.3 and simulated spectra of $[(n\text{-Bu}_3P)_2Ru(C_6H_6NO)(C_7H_{10}N_2)]^+$.

Experimental part

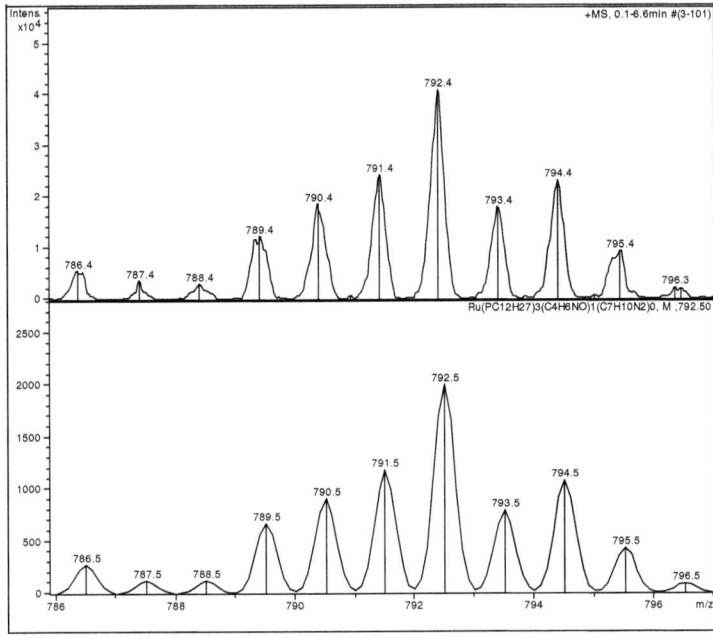

m/z (% relative to peak at 792.4) = 786.4 (14), 787.4 (8), 788.4 (7), 789.4 (29), 790.4 (45), 791.4 (58), 792.4 (100), 793.4 (43), 794.4 (52), 795.4 (22), 796.3 (6), Calc.: 786.5 (14), 787.5 (6), 788.6 (6), 789.5 (34), 790.5 (46), 791.5 (59), 792.5 (100), 793.5 (39), 794.5 (54), 795.5 (22), 796.5 (5).

Figure 11: amplified area of the signal at m/z = 792.4 and simulated spectra of $[(n\text{-Bu}_3P)_3Ru(C_6H_6NO)]^+$.

Experimental part

5.7.2 Procedure for the in situ NMR

A 5 mm NMR tube was charged with [Ru(cod)(met)$_2$] (12.8 mg, 0.04 mmol) and DMAP (9.8 mg, 0.08 mmol). The tube was sealed with a rubber septum and purged with alternating vacuum and nitrogen cycles. Tri-*n*-butylphosphine (30 µL, 0.012 mmol) and toluene-d_8 (1 mL) were added *via* syringe and placed in an ultrasonic bath for 3 min to give a clear yellow solution. Spectra 1h/1p (Figure 12 and Figure 13) were recorded at 25 °C. The sample was heated to 100 °C for 5 min. At 25 °C, spectra 2h/2p (Figure 14 and Figure 15) were recorded. Pyrrolidin-2-one (6.2 µL, 0.08 mmol) was added *via* syringe, the mixture was shaken and spectra 3h/3p (Figure 16 and Figure 17) were recorded at 100 °C.

Spectra 4h/4p (Figure 18 and Figure 19) are the reference spectra ([Ru(cod)(met)$_2$] (12.8 mg, 0.04 mmol) and tri-*n*-butylphosphine (30 µL, 0.012 mmol) in toluene-d_8 (1 mL)) at 25 °C.

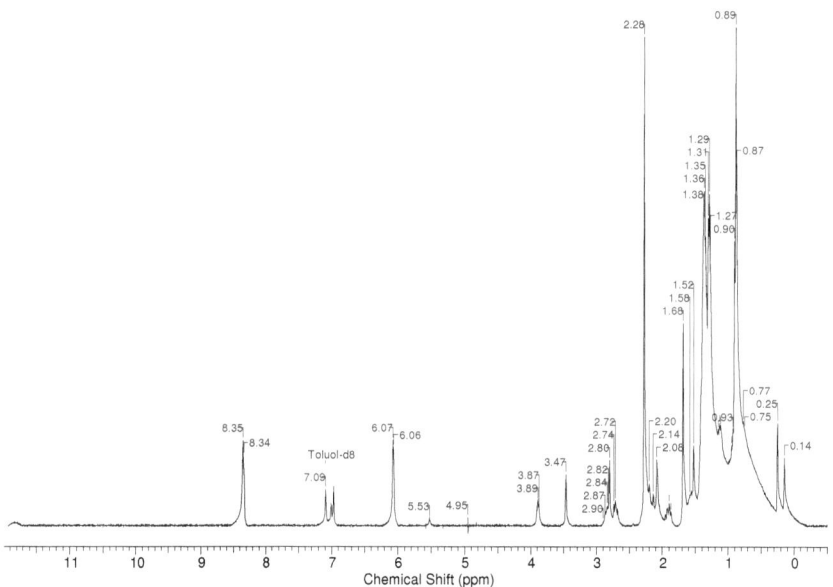

Figure 12: spectrum 1h: ^{1}H-NMR (400.1 MHz, toluene-d_8).

Experimental part

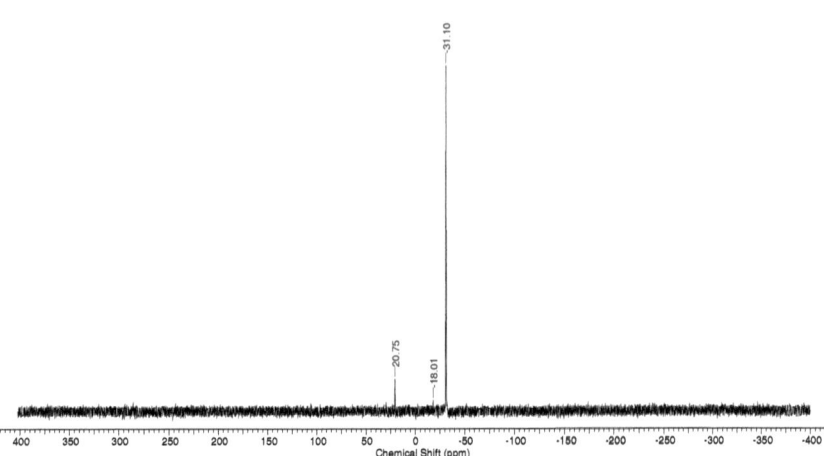

Figure 13: spectrum 1p: ^{31}P-NMR (162.0 MHz, toluene-d_8).

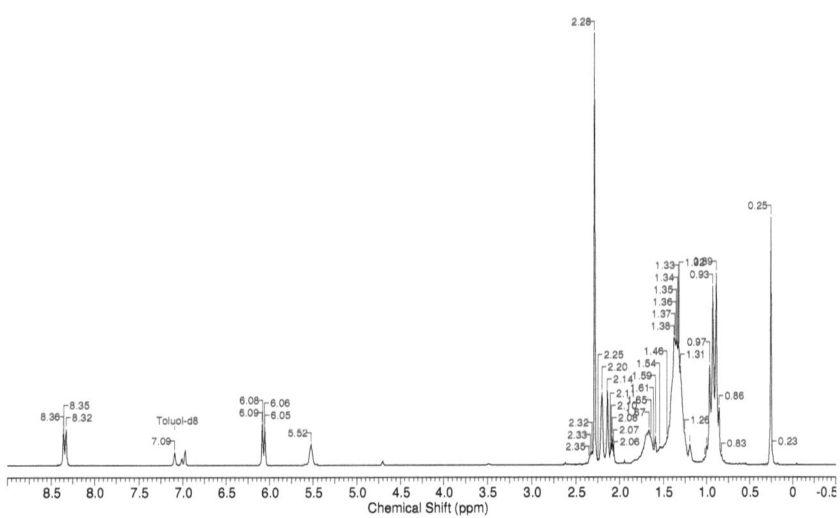

Figure 14: spectrum 2h: ^{1}H-NMR (200.1 MHz, toluene-d_8).

Experimental part

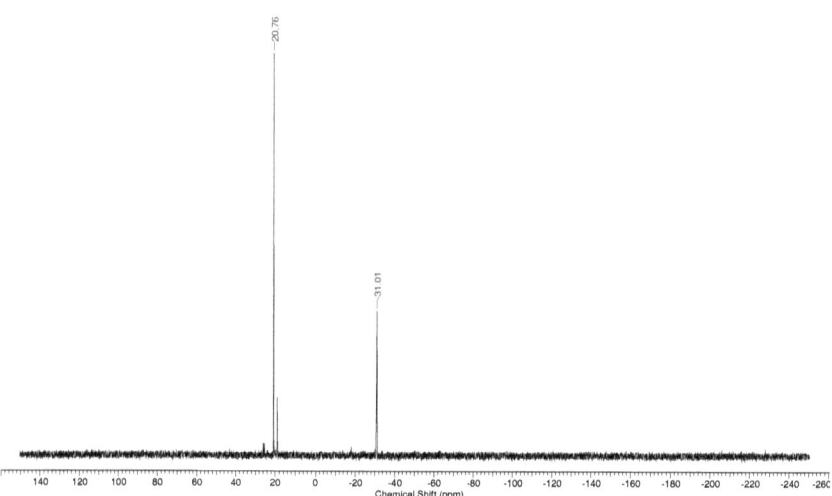

Figure 15: spectrum 2p: ^{31}P-NMR (81.0 MHz, toluene-d_8).

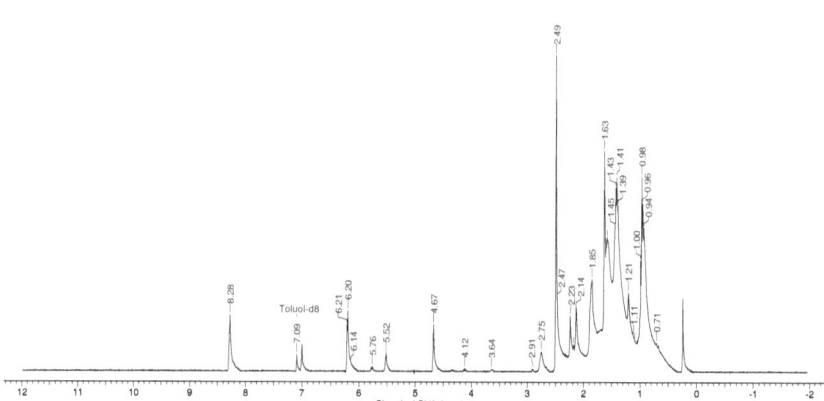

Figure 16: spectrum 3h: ^{1}H-NMR (400.1 MHz, toluene-d_8).

Experimental part

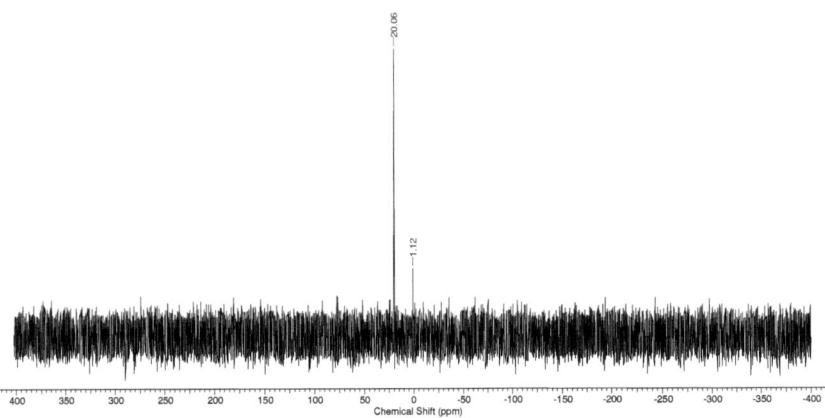

Figure 17: spectrum 3p: ^{31}P-NMR (162.0 MHz, toluene-d_8).

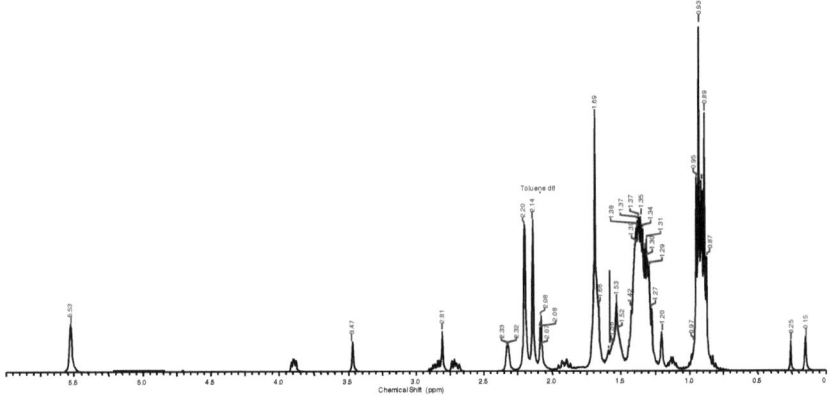

Figure 18: spectrum 4h: ^{1}H-NMR (400.1 MHz, toluene-d_8).

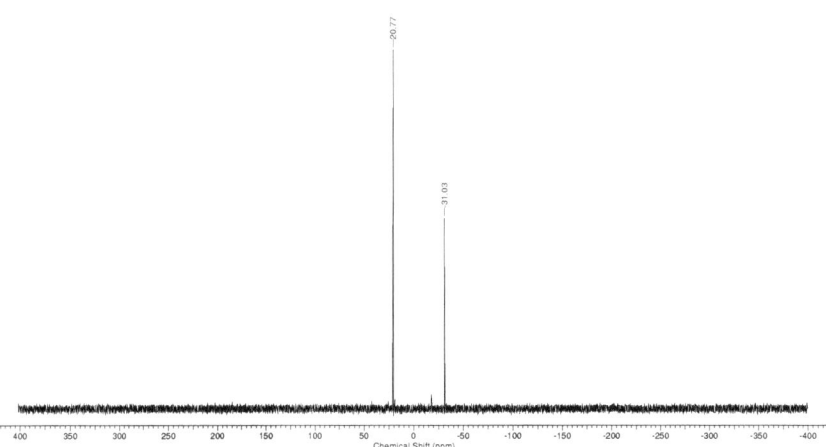

Figure 19: spectrum 4p: ^{31}P-NMR (162.0 MHz, toluene-d_8).

5.7.3 Procedure for the D-experiments

Synthesis of 1-deutero-1-hexyne

In a flame-dried, nitrogen-flushed flask, a small amount of sodium hydride (60% in mineral oil) (28 mg, 70 mmol) was added while stirring to 99% 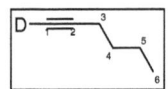 D_2O (20 mL, 1 mol). 1-Hexyne (82 mmol, 9.4 mL) was added, and the two-phase mixture was stirred for 7 h. The operation was repetead. The layers were separated under a nitrogen atmosphere. The organic layer was dried over molecular sieves, filtered, and distilled to afford 1-deutero-1-hexyne (95%).

^{1}H-NMR (400 MHz, CDCl$_3$, 25 °C): δ=2.15 (t, ^3J(H,H)=7.0 Hz, 2H,), 1.32-1.55 (m, 4H, H), 0.88 (t, ^3J(H,H)=7.0 Hz, 3H, H) ppm.

^{13}C-NMR (101 MHz, CDCl$_3$, 25 °C): δ=84.0 (t, ^3J(C,D)=7.40 Hz,) 67.7 (t, ^2J(C,D)=38 Hz,), 30.6, 21.8, 18.0, 13.4 ppm.

Syntheses of N-[1'-deutero-(Z)-hex-1-en-1-yl]succinimide (3aa*)

An oven-dried flask was charged with [Ru(cod)(met)$_2$] (6.4 mg, 0.02 mmol), scandium triflate (19.7 mg, 0.04 mmol) and succinimide (99.1 mg, 1.00 mmol) and flushed with argon. Subsequently, tri-n-butylphosphine (15 μL, 0.06 mmol), 1-deutero-1-hexyne (229 μL, 2.0 mmol), and dry DMF (3.0 mL) were added via syringe. The resulting colorless solution was 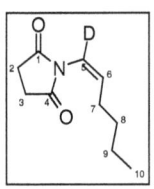 stirred for 15 h at 60 °C, the volatiles were removed in vacuo. The residue was purified by column chromatography (silica gel, ethyl acetate/hexanes 1:3) to afford the title product (131.2 mg, 72 %) as yellowish oil.

^{1}H-NMR (400 MHz, CDCl$_3$, 25 °C): δ=5.66 (t, ^3J(H,H)=7.3 Hz, 1H, H-5), 2.72 (s, 4H, H-2, 3), 1.88 (dt, ^3J(H,H)=7.4, 7,.2 Hz, 2H, H-7), 1.16-1.40 (m, 4H, H-8, 9), 0.81 (t, ^3J(H,H)=7.2 Hz, 3H, H-10) ppm.

^{13}C-NMR (101 MHz, CDCl$_3$, 25 °C): δ=175.4 (C-1, 2), 133.1 (C-6), 116.2 (t, ^3J(C,D)=27 Hz, C-5), 30.5 (C-7), 28.2 (C-2, 3), 27.5 (C-8), 22.1 (C-9), 13.8 (C-10) ppm.

D-NMR (61 MHz, CHCl$_3$ und CDCl$_3$): δ=5,93 (s, D) ppm.

Experimental part

Syntheses of *N*-[1'-deutero-(Z)-hex-1-en-1-yl]pyrrolidinone (14aa*)

An oven-dried flask was charged with [Ru(cod)(met)₂] (6.4 mg, 0.02 mmol) and DMAP (4,90 mg, 0,04 mmol) and flushed with argon. Subsequently, tri-*n*-butylphosphine (15 µL, 0.06 mmol), 1-deutero-1-hexyne (229 µL, 2.0 mmol), and dry toluene (3.0 mL) were added via syringe. The resulting colorless solution was stirred for 15 h at 100 °C, the volatiles were removed in vacuo. The residue was purified by column chromatography (silica gel, ethyl acetate/hexanes 1:3) to afford the title product (129.5 mg, 77 %) as yellowish oil.

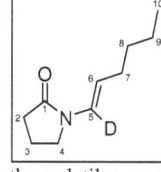

¹H-NMR (400 MHz, CDCl₃, 25 °C): δ=4.83 (m, 1H, H-6), 3.43 (m, 2H, H-4), 2.37 (m, 2H, H-2), 2.06 (m, 4H, H-3, 7), 1.27 (m, 4H, H-8, 9), 0.83 (m, 3H, H-10) ppm.

¹³C-NMR (101 MHz, CDCl₃, 25 °C): δ=172.5 (C-1), 123.5 (t, ³*J*(C,D)=27 Hz, C-5), 112.1 (C-6), 45.1 (C-4), 32.1 (C-7), 31.1 (C-2), 30.1 (C-8), 21.9 (C-9), 18.4 (C-3), 13.7 (C-10) ppm.

D-NMR (61 MHz, CHCl₃ und CDCl₃): δ=6.45 (s, D) ppm.

Syntheses of *N*-[1'-deutero-(Z)-hex-1-en-1-yl]benzamide (9aa*)

An oven-dried flask was charged with [Ru(cod)(met)₂] (6.4 mg, 0.02 mmol) and dcypb (54,1 mg, 0,12 mmol), Yb(OTf)₃ (24,8 mg, 0,04 mmol) and beanzamide (121.1 mg, 1 mmol) and flushed with argon. Subsequently, 1-deutero-1-hexyne (229 µL, 2.0 mmol), and dry DMF (3.0 mL) were added via syringe. The resulting colorless solution was stirred for 15 h at 60 °C, the volatiles were removed in vacuo. The residue was purified by column chromatography (silica gel, ethyl acetate/hexanes 1:3) to afford the title product (161.4 mg, 79 %) as yellowish oil.

¹H-NMR (400 MHz, CDCl₃, 25 °C): δ=7.90 (s, 1H, N-H), 7.74 (d, ³*J*(H,H)=7.7 Hz, 2H, H_{arom}), 7.44 (t, ³*J*(H,H)=7.4 Hz, 1H, H_{arom}), 7.38 (m, 2H, H_{arom}), 4.80 (t, ³*J*(H,H)=7.3 Hz, 1H, H-7), 2.07 (q, ³*J*(H,H)=7.3 Hz, 2H, H-8), 1.37 (m, 2H, H-9), 1.31 (m, 2H, H-10), 0.85 (m, 3H, H-11) ppm.

Experimental part

^{13}C-NMR (101 MHz, CDCl$_3$, 25 °C): δ=164.5 (C$_{arom}$), 133.7 (C$_{arom}$), 131.3 (C$_{arom}$), 128.2 (C$_{arom}$), 126.9 (C$_{arom}$), 120.7 (t, ^3J(C,D)=29 Hz, C-6), 112,4 (C-7), 31.3 (C-8), 25.3 (C-9), 22.1 (C-10), 13.8 (C-11) ppm.

D-NMR (61 MHz, CHCl$_3$ und CDCl$_3$): δ=7,00 (s, D) ppm.

6 Literature

1 G. Brundtland, *Our Common Future*, Oxford University Press, Oxford, **1987**.

2 a) P. T. Anastas, J. C. Warner, *Green Chemistry: Theory and Practice*, Oxford University Press, Oxford, **1998**. b) M. Poliakoff, J. M. Fitzpatrick, T. R. Farren, P. T. Anastas, *Sciences*, **2002**, *297*, 807-810.

3 S. G. Toske, P. R. Jensen, C. A. Kaufman, W. Fenical, *Tetrahedron* **1998**, *54*, 13459-13466.

4 Y. N. Han, G.-Y. Kim, H. K. Han, B. H. Han, *Arch. Pharmacol. Res.* **1993**, *16*, 289-293.

5 L. Yet, *Chem. Rev.* **2003**, *103*, 4283-4306.

6 a) J.-H. Lin, *Phytochemistry* **1989**, *28*, 621-622; b) I. Stefanuti, S. A. Smith, R. J. K. Taylor, *Tetrahedron Lett.* **2000**, *41*, 3735-3738; c) A. Fürstner, C. Brehm, Y. Cancho-Grande, *Org. Lett.* **2001**, *3*, 3955-3957.

7 a) P. J. Stevenson, I. Graham, *ARKIVOC* **2003**, *7*, 139-144. b) C. Gaulon, R. Dhal, T. Chapin, V. Maisonneuve, G. Dujardin, *J. Org. Chem.* **2004**, *69*, 4192–4202.

8 G. J. Roff, Richard C. Lloyd, N. J. Turner, *J. Am. Chem. Soc.* **2008**, *126*, 4098-4099.

9 C. E. Willans, J. M. C. A. Mulders, J. G. de Vries, A. H. M. de Vries, *J. Organomet. Chem.* **2003**, *687*, 494-497.

10 R. Matsubara, Y. Nakamura, S. Kobayashi, *Angew. Chem.* **2004**, *116*, 1711–1713; *Angew. Chem., Int. Ed.* **2004**, *43*, 1679-1681.

11 M. van den Berg, A. J. Minnaard, R. M. Haak, M. Leeman, E. P. Schudde, A. Meetsma, B. L. Feringa, H. M. de Vries, E. P. Maljaars, C. E. Wilians, D. Hyett, A. F. Boogers, J. W. Hendricks, J. G. de Vries, *Adv. Synth. Catal.* **2003**, *345*, 308-323.

12 a) P. Dupau, P. Le Gendre, C. Bruneau, P. H. Dixneuf, *Synlett* **1999**, *11*, 1832-1834. b) X. Wang, J. A. Porco, Jr. *J. Org. Chem.* **2001**, *66*, 8215-8221. c) A. Bayer, M. E. Maier, *Tetrahedron* **2004**, *60*, 6665-6677. d) W. Adam, S. G. Bosio, N. J. Turro, B. T. Wolff, J.

Org. Chem. **2004**, *69*, 1704-1715. e) M. J. Burk, G. Casy, N. B. Johnson, *J. Org. Chem.* **1998**, *63*, 6084-6085. f) C. A. Zezza, M. Smith, *Synth. Commun.* **1987**, *6*, 729-740.

13 a) T. Hosokawa, M. Takano, Y. Kuroki, S.-I. Murahashi, *Tetrahedron Lett.* **1992**, *33*, 6643-6646. b) J. M. Lee, D.-S. Ahn, D. Y. Jung, J. Lee, Y. Do, S. K. Kim, S. Chang, *J. Am. Chem. Soc.* **2006**, *128*, 12954-12962.

14 a) D. J. Wallace, D. J. Klauber, C.-Y Chen, R. P. Volante, *Org. Lett.* **2003**, *5*, 4749-4752. b) X. Pan, Q. Cai, D. Ma, *Org. Lett.* **2004**, *6*, 1809-1812. c) J. L. Brice, J. E. Meerdink, S. S. Stahl, *Org. Lett.* **2004**, *6*, 1845-1848. d) Y. Bolshan, R.A. Batey, *Angew. Chem.* **2008**, *120*, 2139-2142; *Angew. Chem., Int. Ed.* **2008**, *47*, 2109-2112.

15 a) C. B. Ziegler, Jr., R. F. Heck, *J. Org. Chem.* **1978**, *43*, 2949-2952. b) C. A. Busacca, R. E. Johnson, J. Swestock, *J. Org. Chem.* **1993**, *58*, 3299-3203.

16 A. Fürstner, C. Brehm, Y. Cancho-Grande, *Org. Lett.* **2001**, *3*, 3955-3957.

17 S. Krompiec, M. Pigulla, N. Kuźnik, M. Krompiec, B. Marciniec, D. Chadyniak, J. Kasperczyk, *J. Mol. Catal. A: Chem.* **2005**, *225*, 91-101.

18 a) R. Brettle, A. J. Mosedale, *J. Chem. Soc. Perkin Trans. 1* **1988**, 2185–2195; b) K. Kuramochi, H. Watanabe, T. Kitahara, *Synlett* **2000**, 397–399. c) M. Sato, *J. Org. Chem.* **1961**, *26*, 770–779.

19 H. Tsujita, Y. Ura, S. Matsuki, K. Wada, T.-A. Mitsudo, T. Kondo, *Angew. Chem.* **2007**, *119*, 5252-5255; *Angew. Chem., Int. Ed.* **2007**, *46*, 5160-5163.

20 D. A. Ponomarev S. M. Shevchenko, *J. Chem. Educ.* **2007**, *84*, 1725-1726.

21 A. S Hester; K. Himmler, *Ind. Eng. Chem.* **1959**, *51*, 1424-1429.

22 M. Berthelot *Ann. Chim. Phys.* **1863**, *67*, 52-77.

23 H. Lagermarck, A. Eltekoff, *Chem. Ber.* **1877**, *10*, 637-641.

24 a) M. Kutscheroff, *Chem. Ber.* **1881**, *14*, 1540-1542. b) M. Kutscheroff, *Chem. Ber.* **1884**, *17*, 13-29.

25 C. Palomo, M. Oiarbide, J. M. Aizpurua, A. Gonzàlez, J. M. Garcìa, C. Landa, I. Odriozola, A. Linden, *J. Org. Chem.* **1999**, *64*, 8193-8200.

26 L. Hintermann, A. Labonne *Synthesis*, **2007**, *8*, 1121-1150.

27 a) A. S. Carter, U.S. Patent 1896161, **1933**. b) R. F. Conaway, Wilmington, D. U.S. Patent 1967225, **1934**.

28 R. J Thomas, K. N. Campbell, G. F.J. Hennion, *J. Am. Chem. Soc.* **1938**, *60*, 718-720.

29 G. W. Stacy, R. A. Mikulec, *Org. Synth. Coll. Vol. IV* **1963**, 13-15.

30 W. Hiscox, P. W. Jennings, *Organometallics* **1990**, *9*, 1997-1999.

31 R. O: C. Norman, W. J. E. Parr, C. B. J. Thomas, *Chem. Soc., Perkin Trans. 1* **1976**, 1983-1987.

32 S. Teles, S. Brode, M. Chabanas, *Angew. Chem.* **1998**, *110*, 1475-1475; *Angew. Chem. Int. Ed.* **1998**, *37*, 1415-1418.

33 a) E. Mizushima, K. Sato, T. Hayashi, M. Tanaka, *Angew. Chem.* **2002**, *114*, 4745-4747; *Angew. Chem., Int. Ed.* **2002**, *41*, 4563-4565. b) E. Mizushima, D. Cui, D. C. D. Nath, T. Hayashi, M. Tanaka, *Org. Synth.* **2006**, *83*, 55-60.

34 C. Bianchini, J. A. Casares, M. Peruzzini, A. Romerosa, F. J. Zanobini, *J. Am. Chem. Soc.* **1996**, *118*, 4585-4594.

35 M. Tokunaga, Y. Wakatsuki, *Angew. Chem.* **1998**, *37*, 3024-3027; *Angew. Chem., Int. Ed.* **1998**, *37*, 2867-2869.

36 T. Suzuki, M. Tokunaga, Y. Wakatsuki, *Org. Lett.* **2001**, *3*, 735-737.

37 a) C. L. Mao, C. R. Hauser, *Org. Synth.* **1971**, *51*, 90-93; b) C. Kowalski, X. Creary, A. J. Rollin, M. C. Burke, *J. Org. Chem.* **1978**, *43*, 2601-2608; c) J. Libman, Y. Mazur, *Tetrahedron* **1969**, *25*, 1699-1706.

38 W. Kitching, Z. Rappoport, S. Winstein, W.G. Young, G. *J. Am. Chem. Soc.* **1966**, *88*, 2054-2055.

39 G. F. Hennion, D. B. Killian, T. H. Vaughin, J. A. Nieuwland, *J. Am, Chem. Soc.* **1934**, *56*, 1130-1132.

40 K. Weissermel, H. –J. Arpe, *Industrial Organic Chemistry, 2nd edition*, 226-227.

41 M. Rotem, Y. Shvo, *Organometallics* **1983**, *2*, 1689-1691.

42 a) T. Mitsudo, Y. Hori, Y. Watanabe, *J. Org. Chem.* **1985**, *50*, 1566-1568. b) T. Mitsudo, Y. Hori, Y. Yamakawa, Y. Watanabe, *Tetrahedron Lett.* **1986**, *27*, 2125-2126. c) T. Mitsudo, Y. Hori, Y. Yamakawa, Y. Watanabe, *J. Org. Chem.* **1987**, *52*, 2230-2239. d) K. Sakai, K. Kondo, *Chem. Lett.* **1990**, 1665-1666.

43 a) H. Doucet, J. Höfer, C. Bruneau, P. H. Dixneuf, *Chem. Commun.* **1993**, 850-851. b) H. Doucet, B. Martin-Vanca, C. Bruneau, P. H. Dixneuf, *J. Org. Chem.* **1995**, *60*, 7247-7255. c) H. Doucet, N. Derrien, Z. Kabouche, C. Bruneau, P. H. Dixneuf, *J. Organomet. Chem.* **1997**, *551*, 151-157.

44 L. J. Gooßen, J. Paetzold, D. Koley, *Chem. Commun.* **2003**, 706-707.

45 K. Melis, P. Samulkiewicz, J. Rynkowski, F. Verpoort, *Tetrahedron Lett.* **2002**, *43*, 2713-2716.

46 a) C. Bruneau, P. H. Dixneuf, *Angew. Chem.* **2006**, *118*, 2232-2260; *Angew. Chem., Int. Ed.* **2006**, *45*, 2176-2203. b) L. J. Gooßen, N. Rodriguez, K. Gooßen, *Angew. Chem.* **2008**, *120*, 3144-3164; *Angew. Chem. Int. Ed.* **2008**, *47*, 3100-3120.

47 C. Bianchini, A. Meli, M. Peruzzini, F. Zanobini, C. Bruneau, P. H. Dixneuf, *Organometallics* **1990**, *9*, 1155-1160.

48 H. Nakagawa, Y. Okimoto, S. Sakaguchi, Y. Ishii, *Tetrahedron Lett.* **2003**, *44*, 103-106.

49 R. Hua, X. Tian, *J. Org. Chem.* **2004**, *69*, 5782-5784.

50 a) I.G. Farbeind. A.G. Fr. P. 38072 **1930**. b) C.A., 25, 5434 **1931**.

51 J. A. Loritsch, R. R. Vogt, *J. Am. Chem. Soc.* **1939**, *61*, 1462-1463.

52 J. Barluenga, F. Aznar, R. Liz, R. Rodes, *J. Chem. Soc. Perkin trans 1* **1980**, 2732-2737.

53 J. Barluenga, F. Aznar, *Synthesis* **1977**, 195-196.

54 For a review on base-catalyzed hydroamination reactions, see: J. Seayad, A. Tillack, C. G. Hartung, M. Beller, *Adv. Synth. Catal.* **2002**, *344*, 795-813.

55 Y. Li, T. J. Marks, *Organometallics* **1996**, *15*, 3770-3772.

56 A. Haskel, T. Straub, M. S. Eisen, *Organometallics* **1996**, *15*, 3773-3775.

57 a) P. J. Walsh, A. M. Baranger, R. G. Bergman, *J. Am. Chem. Soc.* **1992**, *114*, 1708-1719. b) F. Pohlki, S. Doye, *Chem. Soc. Rev.* **2003**, *32*, 104–114.

58 a) Y. Fukuda, K. Utimoto, H. Nozaki, *Heterocycles* **1987**, *25*, 297-300. b) C. G. Hartung, A. Tillack, H. Trauthwein, M. Beller, *J. Org. Chem.* **2001**, *66*, 6339-6343. c) I. Kadota, A. Shibuya, L. M. Lutete, Y. Yamamoto, *J. Org. Chem.* **1999**, *64*, 4570-4571.

59 a) T. E. Müller, M. Beller, *Chem. Rev.* **1998**, *98*, 675-704. b) F. Alonso, I. P. Beletskaya, M. Yus, *Chem. Rev* **2004**, *104*, 3079-3159.

60 P. L. McGrane, M. Jensen, T. Livinghouse, *J. Am. Chem. Soc.* **1992**, *114*, 5459-5460.

61 P. L. McGrane, T. Livinghouse, *J. Org. Chem.* **1992**, *57*, 1323-1324.

62 E. Haak, I. Bytschkov, S. Doye, *Angew. Chem.* **1999**, *22*, 3584-3586; *Angew. Chem., Int. Ed.* **1999**, *38*, 3389-3391.

63 E. Haak, H. Siebeneicher, S. Doye, *Org. Lett.* **2000**, *2*, 1935-1937.

64 I. Bytschkov, S. Doye, *Eur. J. Org. Chem.* **2001**, 4411-4418.

65 J. S. Johnson, R. G. Bergman, *J. Am. Chem. Soc.* **2001**, *123*, 2923-2924.

66 A. Heutling, S. Doye, *J. Org. Chem.* **2002**, *67*, 1961-1964.

67 Y. Uchimaru, *Chem. Commun.* **1999**, 1133-1134.

68 M. Tokunaga, M. Eckert, Y. Wakatsuki, *Angew. Chem., Int. Ed.* **1999**, *38*, 3222-3225.

69 G. Lavigne, *Eur. J. Inorg. Chem.* **1999**, 917-930.

70 M. Tokunaga, M. Ota, M.-A. Haga, Y. Wakatsuki, *Tetrahedron Lett.* **2001**, *42*, 3865-3868.

71 P. L. Julian, E. W. Meyer, A. Magnani, W. Cole, *J. Am. Chem. Soc.* **1945**, *67*, 1203-1211.

72 Y. Fukumoto, T. Dohi, H. Masaoka, N. Chatani, S. Murai, *Organometallics* **2002**, *21*, 3845-3847.

73 T. Kondo, A. Tanaka, S. Kotachi, Y. Watanabe, *Chem. Commun.* **1995**, 413-414.

74 L. J. Gooßen, J. E. Rauhaus, G. Deng, *Angew. Chem.* **2005**, *6*, 849-852; *Angew. Chem., Int. Ed.* **2005**, *44*, 4042-4045.

75 B. M. Trost, G. R. Dake, *J. Am. Chem. Soc.* **1997**, *119*, 7595-7596.

76 a) F. G. Bordwell, D. Algrim, *J. Org. Chem.* **1976**, *41*, 2507-2508. b) W. N. Olmstead, F. G. Bordwell, *J. Org. Chem.* **1980**, *45*, 32993305. c) M. J. Bausch, B. David, P. Dobrowolski, V. Prasad, *J. Org. Chem.* **1990**, *55*, 5806-5808. d) F. G. Bordwell, H. E. Fried, *J. Org. Chem.* **1991**, *56*, 4218-4223.

77 a) S. Ghosh, D. B. Datta, N. Sen, *Synth. Commun.* **1987**, *17*, 299-307. b) A. Maxwell, D. Rampersad, *J. Nat. Prod.* **1989**, *52*, 411-414. c) J. C. Estévez, M. C. Villaverde, R. J. Estévez, J. A. Seijas, L. Castedo, *Synth. Commun.* **1990**, *20*, 503-507.

78 J. C. Estévez, R. J. Estévez, L. Castedo, *Tetrahedron* **1995**, *51*, 10801-10810.

79 All prices quoted from Strem Chemicals for Research, Catalog No. 21, 2006–2008.

80 a) W. P. Griffith, *The Chemistry of the Rare Platinum Metals: Os, Ru, Ir, and Rh*, Wiley-Interscience, New York, **1967**. b) F. A. Cotton, G. Wilkinson, C. A. Murillo, M. Bochman, *Advanced Inorganic Chemistry, 6th ed.*, John Wiley & Sons, New York, **1999**. c) S. Komiya, M. Hirano, *Synthesis of Organometallic Compounds*, S. Komiya (Ed.), John Wiley & Sons, New York, **1997**. d) J. C. Bailar, H. J. Emeléus, R. Nyholm, A. F. Trotman-Dickenson, *Comprehensive Inorganic Chemistry, Volume 3*,Pergamon Press, Oxford, **1973**.

81 L. J. Gooßen, M. Arndt, M. Blanchot, Felix Rudolphi, F. Menges, G. Niedner-Schatteburg, *Adv. Synth. Catal.* **2008**, *350*, 2701-2707.

82 F. Micoli, L. Salvi, A. Salvini, P. Frediani, C. Giannelli, *J. Organometal. Chem.* **2005**, *690*, 4867–4877.

83 T. Li, A. J. Lough, R. H. Morris, *Chem. Eur. J.* **2007**, *13*, 3796-3803.

84 a) U. Koelle, J. Kossakowski, *Inorg. Syn.* **1992**, *29*, 225-228. b) U. Koelle, J. Kossakowski, *J. Organomet. Chem.* **1989**, *362*, 383-398.

85 R. Steffen, R. Schöllhorn, *Solid State Ionics* **1986**, *22*, 31-41.

86 M. R. Netherton, G. C. Fu, *Org. Lett.* **2001**, *3*, 4295-4298.

87 C. Amatore, A. Jutand, *Acc. Chem. Res.* **2000**, *33*, 314-321.

88 a) A. Couture, R. Dubiez, A. Lablache-Combier, *Chem. Commun.* **1982**, 842-843. b) A. Couture, R. Dubiez, A. Lablache-Combier, *J. Org. Chem.* **1984**, *49*, 714-717. c) B. Yde, N. M. Yousif, U. Pederson, I. Thomsen, S. –O. Lawesson, *Tetrahedron* **1984**, *40*, 2047-2052.

89 a) A. Couture, R. Dubiez, A. Lablache-Combier, *Tetrahedron* **1984**, *40*, 1835-1844. b) K. Yamamoto, S. Yamazaki, I. Murata, *J. Org. Chem.* **1988**, *53*, 5374-5376.

90 For a *non*-catalytic addition of thioamides to conjugated alkynes, see: A. S. Nakhmanovich, T. E. Glotova, T. N. Komarova, G. G. Skvortsova, M. B. Sigalov, V. B. Modonov, *Zh. Org. Khim.* **1983**, *19*, 1428-1431.

91 a) E. H. Bahaji, P. Bastide, J. Bastide, C. Rubat, P. Tronche, *Eur. J. Med. Chem.* **1988**, *23*, 193-197. b) S. Lacroix, V. Rixhon, J. Marchand-Brynaert, *Synthesis* **2006**, *14*, 2327-2334.

92 Bordwell, F. G.; Fried, H. E. *J. Org. Chem.* **1991**, *56*, 4218-4223.

93 J. Inanaga, Y. Baba, T. Hanamoto, *Chem Lett.* **1993**, 241-244.

94 D. D. Perrin, W. L. F. Armarego, D. R. Perrin, *Purification of Laboratory Chemicals 2nd Ed.* Pergamon Press, Oxford **1980**.

VDM Verlagsservicegesellschaft mbH

Die VDM Verlagsservicegesellschaft sucht für wissenschaftliche Verlage abgeschlossene und herausragende

Dissertationen, Habilitationen, Diplomarbeiten, Master Theses, Magisterarbeiten usw.

für die kostenlose Publikation als Fachbuch.

Sie verfügen über eine Arbeit, die hohen inhaltlichen und formalen Ansprüchen genügt, und haben Interesse an einer honorarvergüteten Publikation?

Dann senden Sie bitte erste Informationen über sich und Ihre Arbeit per Email an *info@vdm-vsg.de*.

Sie erhalten kurzfristig unser Feedback!

VDM Verlagsservicegesellschaft mbH
Dudweiler Landstr. 99   Telefon +49 681 3720 174
D - 66123 Saarbrücken   Fax   +49 681 3720 1749
www.vdm-vsg.de

Die VDM Verlagsservicegesellschaft mbH vertritt

Printed by Books on Demand GmbH, Norderstedt / Germany